Volcanoes, Volcanic Rocks and Earthquakes

Other books by Patrick Nurre: These are all also available with sample rock, mineral, and fossil kits at NorthwestRockAndFossil.com.

Rocks and Minerals for Little Eyes (PreK-3)
Fossils and Dinosaurs for Little Eyes (PreK-3)
Volcanoes for Little Eyes (PreK-3)
Geology for Kids (and Geology Journal) (Grades 4-8)
Rock Identification Made Easy (Grades 3-12)
Rock Identification Field Guide
Fossil Identification Made Easy (Grades 3-12)
Fossil Identification Field Guide
Mineral Identification Made Easy (Grades 5-12)
Bedrock Geology (high school)
Fossils, Dinosaurs and Cave Men (high school)
Rocks and Minerals: The Stuff of the Earth (high school)
The Geology of Yellowstone – A Biblical Guide
Genesis Rock Solid – A Biblical View of Geology
Geology and the Hawaiian Islands (Grades 5-12)
Geology and Our National Parks (Grades 5-12)

Volcanoes, Volcanic Rocks and Earthquakes
A Biblical View

Patrick Nurre

The Northwest Treasures Curriculum Project
Building Faith for a Lifetime of Faith

Volcanoes, Volcanic Rocks and Earthquakes, 2nd edition
Published by Northwest Treasures
Bothell, Washington
425-488-6848
NorthwestRockAndFossil.com
northwestexpedition@msn.com

Contents

How to Use This Study

When it comes to studying Earth history and geology, it is vitally important that we help our students to develop a sound perspective about how this all relates to the Bible and especially to the Great Flood of Noah's time. That is the thrust of the *Northwest Treasures Curriculum Project*, of which this study is one component. ***Bedrock Geology***, which is the heart of this program, helps to set the stage for studying all the varying components of geology, and is a great pre-study for this one. *Volcanoes, Volcanic Rocks and Earthquakes* can, however, stand as a study on its own.

Suggestions for use:

1. This study takes about a semester. A suggestion would be to do this study 3-4 days each week: Day 1, define the *Words to know*, and read the lesson. The words may not be defined in the text. You will need to look up some of them. Day 2, reread lesson, highlighting important concepts. Days 3-4, do activities, and take the quiz.
2. Keep a notebook that has a separate section for vocabulary. Each section of the study has words that are vital to understand. If you encounter other geological terms that you don't know, be sure to include them as well.
3. If it is possible, try to get out into the field to find examples of the rocks you are examining. If you live near a volcanic area, take time for a field trip to research it.

Materials

This study is best used with rock and mineral samples, available from Northwest Treasures (at NorthwestRockAndFossil.com). Samples include:

- Quartz – rock-forming mineral
- Muscovite – rock-forming mineral
- Potassium feldspar – rock-forming mineral
- Sodium feldspar – rock-forming mineral
- Jasper – rock-forming mineral
- Calcite – rock-forming mineral
- Olivine – rock-forming mineral
- Biotite mica – rock-forming mineral
- Magnetite (iron ore) – rock-forming mineral
- Calcium feldspar – rock-forming mineral
- Amphibole – rock-forming mineral
- Pyroxene – rock-forming mineral
- Cinders
- Laccolith magma
- Basalt
- Vesicular basalt
- Pahoehoe
- A'a
- Scoria
- Dark pumice
- Ash
- Tuff
- Amygdaloidal basalt
- Olivine in basalt
- Oxidized basalt
- Basalt porphyry
- Rhyolite
- Perlite
- Light colored pumice
- Obsidian
- Andesite
- Dacite
- Vitrophyre
- Fault gouge

Lesson One – What are Volcanoes?

Mt. Fuji

Words to Know: volcano fissure uniformitarianism volcanic rock interpretation worldview catastrophic magma The Age of Enlightenment (the Enlightenment) accountable mythological the rock cycle

Introduction

One of the confusing things about developing a Biblical view of Earth history is the nomenclature or language of science. For instance, the word ***volcano*** is not mentioned in the Bible. Ever since the ***Enlightenment*** (1700s-1800s) the science community has adopted the language of the Greeks and Romans – Greek and Latin. The Old Testament, written in Hebrew, was written before this adoption and the New Testament was written in Greek that is slightly different from Modern Greek. The modern student rarely studies Greek or Latin and so the language of modern geology can be intimidating and can seem secretive. Consequently, the student of science must learn a new language in order to comprehend modern geology and its relationship to the Bible.

The word ***geology*** comes from two Greek roots, *geo*, the earth, and *logi*, the study of – the study of the earth. This subject is huge, and it includes the following:

a) The study of the present geologic forces and processes that are now at work on the earth
b) The study of the rocks and minerals that are a part of these forces
c) The study of the landforms on the earth, of which volcanoes would be a part
d) The study of Earth History

What is a Volcano?

The word ***volcano*** comes from the name of a ***mythological*** god named Vulcan – the god of fire. And that does aptly illustrate the nature of volcanoes – explosions of fire and hot steam, rock, ash, and lava. But from a Scriptural viewpoint, volcanoes illustrate something much more. Volcanoes are extremely destructive geologic events. In addition to the blast and destructive rolling clouds of bits and pieces of ***volcanic rock*** and very hot steam and ash, there are all kinds of poisonous gasses that are emitted during an eruption. Rolling lava flows can destroy property and life very quickly.

So, volcanoes are ***fissures*** or openings in the earth that release destructive forces to man and his environment.

Volcanoes are actually weaknesses in the earth left over from another time in Earth history. In Genesis chapter 1:31, the Scripture states, "God saw all that He had made, and behold, it was very good. And there was evening and there was morning, the sixth day." How would volcanoes fit into this view of the finished creation?

Discuss: From a Scriptural viewpoint, how could these volcanic menaces be part of the "very good" earth talked about in Genesis chapter one?

History and Worldview

Before we get into the study of volcanoes, we need to clear up a misunderstanding that continues to this day and is taught by both science and some churches. Our public institutions continue to teach that in the debate about origins and Earth history, the issue is *Science vs. Religion or Science vs. the Bible*. This is a huge distortion of the truth. The true issue since the Enlightenment has always been, *Science plus philosophy vs. Science plus the Scriptures*. In our study of geology and in particular, volcanoes, it will be necessary to constantly distinguish between these two conflicting ***worldviews***. The word *science* comes from a Latin word meaning, knowledge. It is not in the area of knowledge where the Scriptures conflict with modern geology, but in the area of worldview; the conflict comes in the ***interpretation*** (worldview) of the knowledge we gain by studying the earth.

In the late 1700s the Father of Modern Geology, James Hutton, a Scottish physician, taught that according to his observations of geologic processes that he witnessed, that he "...could find no vestige of a beginning, no prospect of an end." [1] In other words, the geological formations we see around us have been always changing so that we can't properly see how they started, and this process will continue as long as the earth exists. He came to this conclusion because he assumed that these geologic processes had always operated at the pace he presently observed. He reasoned that the earth must be older, much older than what the Bible stated.

This was the beginning of an idea called ***uniformitarianism*** – *present geologic processes are sufficient to explain the past operations of the earth.* Consequently, no god or ***catastrophes*** like Noah's Flood were needed to explain the past history of the earth and indeed, according to Hutton,

[1] *Theory of the Earth; or an Investigation of the Laws observable in the Composition, Dissolution, and Restoration of Land upon the Globe* and appeared in *Transactions of the Royal Society of Edinburgh*, vol. I, Part II, pp. 209-304, plates I and II, published 1788.

no god should be allowed into the realm of science to explain such natural phenomenon. This sums up the Enlightenment period in Western Civilization – a time of drifting from the church, God and the Scriptures.

In the 1830s another man named Charles Lyell, a British Lawyer, popularized Hutton's ideas in a phrase that is still used today: *"The present is the key to the past."* And so, modern geology has continued to this day to view the past history of the earth in terms of present observable processes.

Here is an example - today we observe an abundance of ***volcanic rocks*** all over the earth. Some layers of these volcanic rocks exceed more than a mile in thickness. Modern geology would interpret these layers as evidence that the origin of the earth was begun in volcanic eruptions and then slowly changed over millions of years into the soils and landscapes we see today. A Scriptural viewpoint would interpret these layers as a result of the Genesis catastrophic Flood and its aftermath. Why the conflict? Is it because the Bible teaches scientific error? There is nothing in the science of the volcanic rocks that would tell us that they are millions and billions of years old. But it is the worldview, the interpretation, that would determine this. (We will discuss more about this in ***Lesson Six***.)

In Genesis 7:11 we read, *"In the six hundredth year of Noah's life, in the second month, on the seventeenth day of the month, on the same day all the fountains of the great deep burst open, and the floodgates of the sky were opened."* This historical account records the horrendous geologic events of the first day of the Genesis Flood – *"...the fountains of the great deep burst open...."* This event would certainly imply that there must have been fissures opened in the earth through which came an abundance of water and ***magma***.

So the volcanic rocks which cover the earth were the result of the different stages of the Genesis Flood and its catastrophic upheaval. Volcanic eruptions continue to this day, but with much less frequency and intensity. Volcanoes are ruptures; weaknesses in a once very good Earth. Volcanic eruptions serve as a reminder that God held man ***accountable*** for his moral choices and that there will be a day in the future when He will once again hold man accountable.

The origin of volcanoes as interpreted from a Biblical viewpoint is not some mythological story to explain the wrath and anger of the gods, but a real historical event brought on by God as a consequence of man's rebellion against his creator. Ever since the Enlightenment, the Scriptural history of the global Flood of Noah's day has been relegated to the domain of myth. The idea of God's involvement in His creation has been called nonsense by secularists. If the Genesis Flood was indeed a consequence of man's rebellion, then man must recognize a God Who rules over all, and that calls man into accountability for his actions.

Memorize: Genesis 7:11

The Rock Cycle

James Hutton viewed the history of the earth in a cyclical way. He saw the geologic history of the earth as sort of giant recycling process, constantly reusing the materials of the earth to generate new crust, rocks, and change in existing landforms. Volcanoes represented part of this ongoing history. Of course, if we did not have Biblical revelation to tell us about the origin of the earth and other significant geologic events like the global Flood, then it might be reasonable to ponder explanations like Hutton's. But we do have these historical events recorded and authenticated by Jesus in the New Testament (Matthew 24:37-39, Matthew 19:4). To ignore them would truly lead to a blindness of God's involvement in His creation.

Today, Hutton's views have been expressed in a popular diagram called the ***rock cycle***. What is the rock cycle? It is a diagram taught as pure science by our public institutions and many private and Christian schools. It portrays the idea that the earth's rocks recycle. That is, rocks are constantly being reproduced in the Earth's mantle and core, eroded and changed by erosive and metamorphic process to be used again and again – a sort of giant recycling machine. No wonder Hutton said he could see no vestiges of a beginning nor prospect of an end! Is this representation true?

Study the typical illustrations of the rock cycle below and record what you observe. Do you think these processes are always true? When might they have been different? Have these processes been observed?

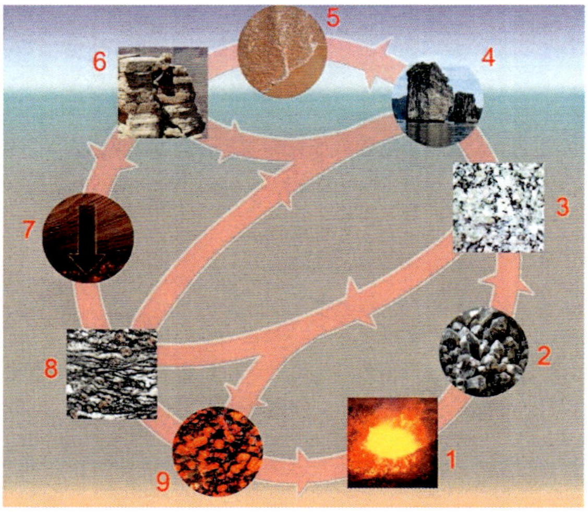

A diagram of the rock cycle. Legend: 1. magma; 2. crystallization (freezing of rock); 3. igneous rocks; 4. erosion; 5. sedimentation; 6. sediments and sedimentary rocks; 7. tectonic burial and metamorphism; 8. metamorphic rocks; 9. melting.

Now, there is no doubt that some of these processes might occur under certain circumstances. But geologists teach these processes as if they were totally true. They also teach them as processes that were slow and gradual over hundreds of millions of years. But to say that these processes represent all of Earth history is an assumption that cannot be scientifically proven. If the geologic events described in Genesis are historical events, then processes would have been totally different during these particular events. The Genesis Flood describes an event that involved rapid erosion, rapid deposition, rapid transportation, and rapid rock formation. So, what is the science in volcanoes and volcanic eruptions and how does the Bible explain such horrific events? We'll look at the answer to this question in the next section.

Activity #1

Charting the History of the Enlightenment and its Influence on the Rejection of a Biblical View of Earth History

Research the historical period called the Enlightenment. Record the main thinkers and events that led up to a rejection of the Biblical view of the Genesis Flood. Construct a chart that illustrates these events from the 18th through the 19th centuries. Ideas for investigation might include influential thinkers, the French Revolution, the American Revolution, Darwin's journey on the Beagle, etc.

Take Quiz 1 (back of book).

Lesson Two – Types of Volcanoes

Words to know: fissure volcano   shield volcano   dome volcano   cinder cone composite volcano   stratovolcano   caldera   scoria   volcanic vent   basalt   butte Ring of Fire   oceanic trench   rift   laccolith   planation surface   intrusion   sheet erosion channelized erosion

Types of Volcanoes

Not all volcanoes are alike. Learning the various features of the different volcanoes will help you identify different volcanic rocks in a later section. Let's take a look at some of the various types of volcanoes. (Although different types of volcanic rocks are necessarily mentioned in this lesson, we will wait until a later lesson to discuss them.)

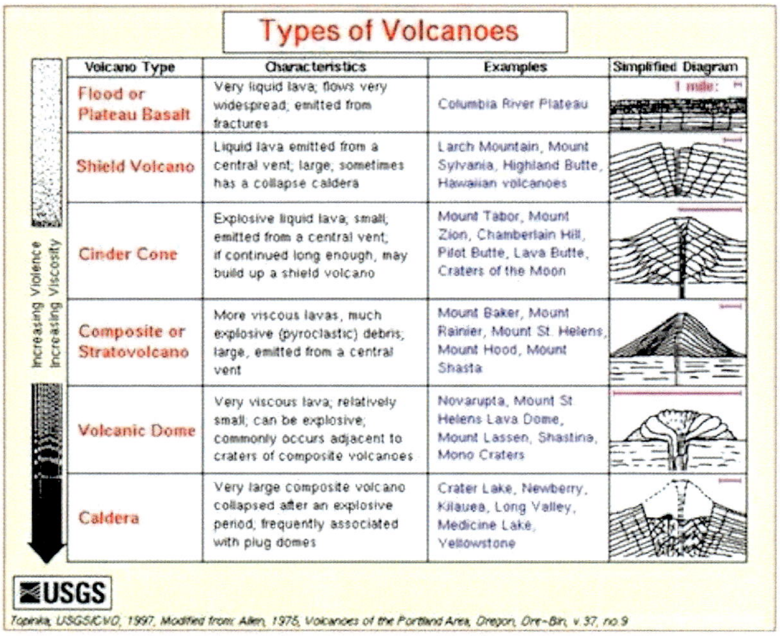

Types of Volcanoes

Volcano Type	Characteristics	Examples	Simplified Diagram
Flood or Plateau Basalt	Very liquid lava; flows very widespread; emitted from fractures	Columbia River Plateau	
Shield Volcano	Liquid lava emitted from a central vent; large; sometimes has a collapse caldera	Larch Mountain, Mount Sylvania, Highland Butte, Hawaiian volcanoes	
Cinder Cone	Explosive liquid lava; small; emitted from a central vent; if continued long enough, may build up a shield volcano	Mount Tabor, Mount Zion, Chamberlain Hill, Pilot Butte, Lava Butte, Craters of the Moon	
Composite or Stratovolcano	More viscous lavas, much explosive (pyroclastic) debris; large, emitted from a central vent	Mount Baker, Mount Rainier, Mount St. Helens, Mount Hood, Mount Shasta	
Volcanic Dome	Very viscous lava; relatively small; can be explosive; commonly occurs adjacent to craters of composite volcanoes	Novarupta, Mount St Helens Lava Dome, Mount Lassen, Shastina, Mono Craters	
Caldera	Very large composite volcano collapsed after an explosive period; frequently associated with plug domes	Crater Lake, Newberry, Kilauea, Long Valley, Medicine Lake, Yellowstone	

Fissure volcanoes (also called flood or plateau ***basalt*** flows); These are vents, or ***rifts*** through which lava erupts, usually without any explosive activity. Fissure vents can cause large basalt floods (or flows) and lava channels. This type of volcano is usually hard to recognize from the ground and from outer space because it has no central crater or opening, and the surface is mostly flat. The volcano can usually be seen as a crack in the ground or on the ocean floor. An example of a basalt flood as a result of a fissure volcano would be the Columbia River Plateau.

Shield volcanoes look like a huge shield, convex side up. The lava comes through a central vent. Shield volcanoes usually stretch for miles and are not easily visible close up. Shield volcanoes are built almost entirely of fluid basalt lava flows. Examples would be the Hawaiian volcanoes, Mauna Kea and Mauna Loa.

Mauna Loa is a shield volcano on the Big Island of Hawaii.

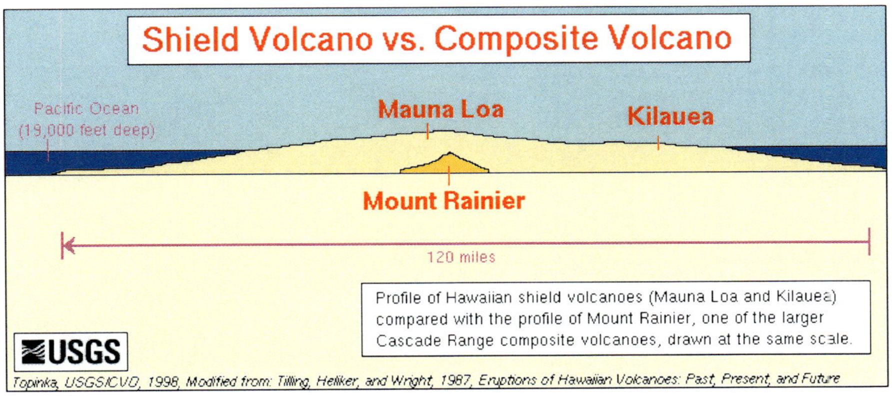

Mauna Loa is the earth's largest shield volcano.

Ash or cinder cones usually erupt just once in their life cycle. Another name for a cinder cone is ***scoria cone*** because the type of lava it erupts is the volcanic rock scoria, which is airy, lightweight, basaltic lava. The lava is explosive and liquid, emitted through a central vent. Examples include Craters of the Moon, SP Crater, and Lava ***Butte***.

Craters of the Moon, Idaho; SP Crater in the San Francisco volcanic field of Arizona is an example of an ash/cinder cone volcano.

This cinder cone is located on Mauna Kea shield volcano, Hawaii.

Composite or stratovolcanoes build high cones made of repeated volcanic eruptions and lava flows. These lavas are more viscous and explosive and have a large central vent. These can build up very quickly. Examples include Mt. Fuji, Mount Rainier, Mt. Hood, and Mt. St. Helens. Mt. St. Helens is both a dome and a stratovolcano.

Mt. Fuji is a composite or stratovolcano; Mt. Hood is a composite or stratovolcano.

Dome volcanoes often form inside of the craters of existing composite or stratovolcanoes, or adjacent to the volcano. Their lava is very viscous, rhyolite or dacite lava. They tend to be smaller volcanoes. Examples would include the Mt. St. Helens Dome, Shastina, and Chaitén.

Chaitén, near the Gulf of Corcovado in southern Chile, South America, is a dome volcano.

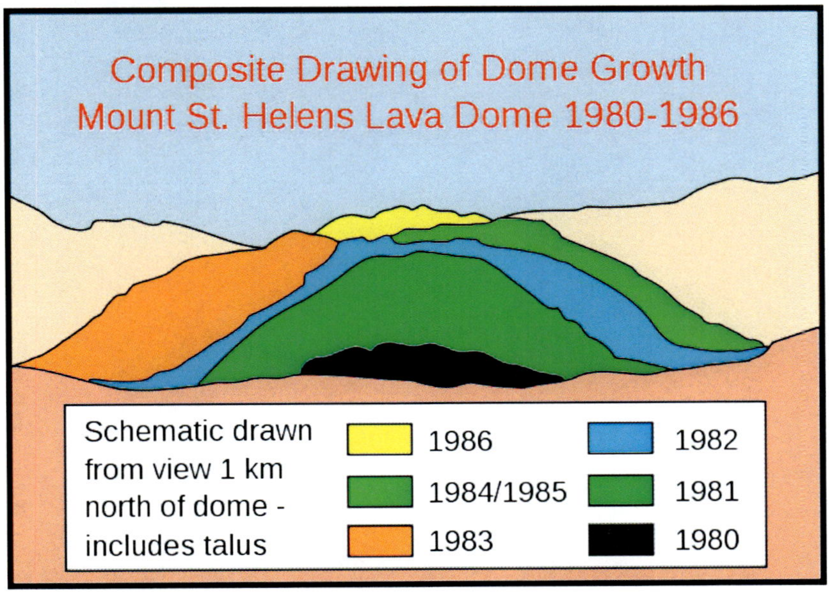

Mt. St. Helens in Washington state is a dome volcano.

360 view of Mt. St. Helens with new dome visible in the center

Calderas are cauldron-like volcanic features usually formed by the collapse of land following a volcanic eruption. The word comes from the Spanish word, *caldera,* which comes from the Latin, *calderia,* meaning cooking pot. The term *cauldron* is used in some English texts. Calderas are sometimes confused with volcanic craters.

Examples of calderas include Crater Lake, Newberry, Kilauea, and Yellowstone. These calderas are so large that they are not easily noticed. It took scientists a couple of hundred years to recognize them as calderas. The advent of aerial photography helped to see these immense catastrophic landforms.

An example of a caldera volcano (super volcano) is the Yellowstone Caldera in Wyoming.
The Yellowstone Caldera is approximately 45 miles long and 30 miles wide – truly a huge catastrophic and destructive volcano! The Yellowstone Caldera is considered to have been the largest volcanic eruption in Earth's history. What event could have triggered such huge eruptions in our past history? Were these just part of the on-going recycling geologic process envisioned by James Hutton? Or were they part of the Genesis catastrophic flood that destroyed God's creation as a result of man's rebellion?

The blue outline shows the approximate area of the Yellowstone Caldera in Yellowstone National Park.

Laccoliths

The word comes from two Greek roots, meaning *pond stone*. A laccolith is technically a pool or pond of magma that did not quite erupt through the earth's surface to become a volcano and so formed a hill or mound usually in sedimentary rock. In other words, the rise of magma never quite became a volcano. Laccoliths typically consist of relatively viscous magmas, high in quartz. The rock formed looks very much like either andesite lava, an intermediate volcanic rock, or gray diorite, an intermediate plutonic rock.

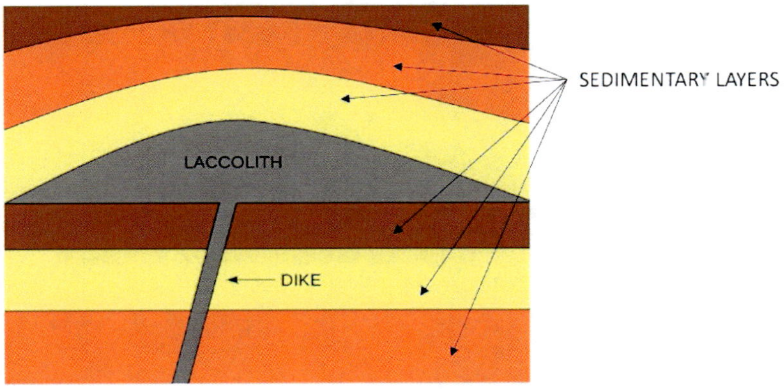

Laccoliths are usually inconspicuous, as they are generally covered over by grass, shrubs, and trees, and could be mistaken for just an odd hill. A person can drive by laccoliths and not even be aware of them. But once you recognize one, you will never forget it.

A laccolith in central Montana

The following picture is of a laccolith (Round Mountain) which is part of the La Sal Mountains in Southeastern Utah. It is located in a sandstone canyon a few miles outside of Moab, Utah. It is a perfect illustration of an ***intrusion*** within sedimentary rock, having never quite erupted through, with the softer sandstone washed away during the sheet erosion stage of the Flood.

As a laccolith is formed under sedimentary layers, we naturally want to know how they surfaced. Exposure of the laccoliths would have been accomplished during one of the stages of the Genesis Flood. Chapter 8 of Genesis states that during the first 150 days, the entire earth was inundated with water. This could be called the first stage of the Flood. The second stage of the Flood would have occupied the next 150-175 days. We can divide this stage into a sheet flow (or ***sheet erosion***) stage, where massive amounts of water ran off of the rising mountains in sheets according to Psalm 104:5-9, and then the channelized flow (or ***channelized erosion***) stage of the Flood, where deep ravines were carved by retreating Flood waters. These ravines are still visible in mountains today.

Lone Mountain near Bozeman, Montana, is a laccolith.

First Stage of the Genesis Flood

During the first stage of the Genesis Flood, the entire earth would have been covered with water. Thousands of feet of sediments would have been deposited during this stage as tectonic and volcanic activity would have continued. Laccoliths would have been actively pushing into these sediments forming domes of magma which then hardened.

Second Stage of the Genesis Flood

With the start of the second stage of the Genesis Flood, huge amounts of still pliable sediments would have been sheared and channelized by massive amounts of water moving off of the earth into the freshly formed ocean basins. This is the idea expressed in Psalm 104:5-9. It is during the second stage of the Genesis Flood that ***planation surfaces*** and canyons would have been excavated rapidly. These kinds of formations could not have been formed gradually and over long periods of time. This is where modern geology has made some serious mistakes. They rejected the Genesis Flood as a part of history and therefore mistakenly interpreted these geological formations.

The following picture is of Square Butte east of Great Falls, Montana. It is a laccolith with a planation surface. A planation surface is a surface that has the appearance of being sheared flat through some kind of rapid erosion event. I say rapid because planation surfaces are uniform in appearance. Their erosion was not gradual, or they would have shown an uneven erosion surface. The second stage of the Flood would have been the perfect mechanism to explain planation surfaces.

Planation surfaces are some of the biggest mysteries in modern secular geology today. They remain an anomaly, something that cannot be explained by modern geology. A uniformitarian explanation of this area says, "Erosion has worn away hundreds of feet of overlying sedimentary rock, leaving remnants of the more resistant igneous formations."[2] While we would not disagree with this explanation in principle, uniformitarianism requires millions of years for this to have happened. It is the small and slow increments of erosion that they would claim that do not fit with the evidence.

[2] "The Crown Butte Story," quote found at http://formontana.net?Z.html, web site by Rod Benson, earth science teacher at Helena High School, Helena, Montana.

Some think that Devil's Tower in Wyoming is a laccolith. Others think it is simply a volcanic neck. Devil's Tower rests on a sedimentary base. What happened to the rest of the sedimentary rock? It is clear that Devil's Tower must have been covered by softer sedimentary sediments. What happened to them? The second stage of the Flood might have acted as a mechanism for removing the surrounding softer sedimentary sediments, having been rapidly eroded away by the receding Flood waters. Whether this happened as a result of millions of years or from the Flood is a question that cannot be answered by science, but only by one's worldview.

Devil's Tower as seen in 1900 and today.

The following picture is of the La Sal Mountains in southeastern Utah. Remnants of the sedimentary layers in the foreground were originally laid down by the Flood. The La Sal Mountains in the distance are huge laccoliths. You can get an idea of just how thick the sediments must have been. The La Sal Mountains were originally below sedimentary layers hundreds of feet thick. According to a Genesis Flood view, as the La Sals rose up, the sedimentary layers would have been sheared and washed off of the surface in the sheet flow stage of the Genesis Flood, exposing the La Sal Mountains. You can see evidence of this in the planation surface of the sedimentary deposits in the foreground. It is at the 8,500-foot level of these mountains that well-preserved theropod dinosaur tracks are found in sandstone.

The Ring of Fire

The Ring of Fire is an area where a large number of volcanic eruptions and earthquakes occur in the basin of the Pacific Ocean. It is a 25,000 square mile horseshoe-shaped arc from Australia to South America, and is the most active volcanic activity chain in the world. A volcanic arc is a chain of volcanoes positioned in an arc shape, as you can see on the following map. The Ring of Fire has

452 volcanoes and is home to over 75% of the world's active and dormant volcanoes as well as thousands of earthquakes each year. About 90% of the world's earthquakes and 81% of the world's largest earthquakes occur along the Ring of Fire. On the following map, each red dot represents an active volcano.

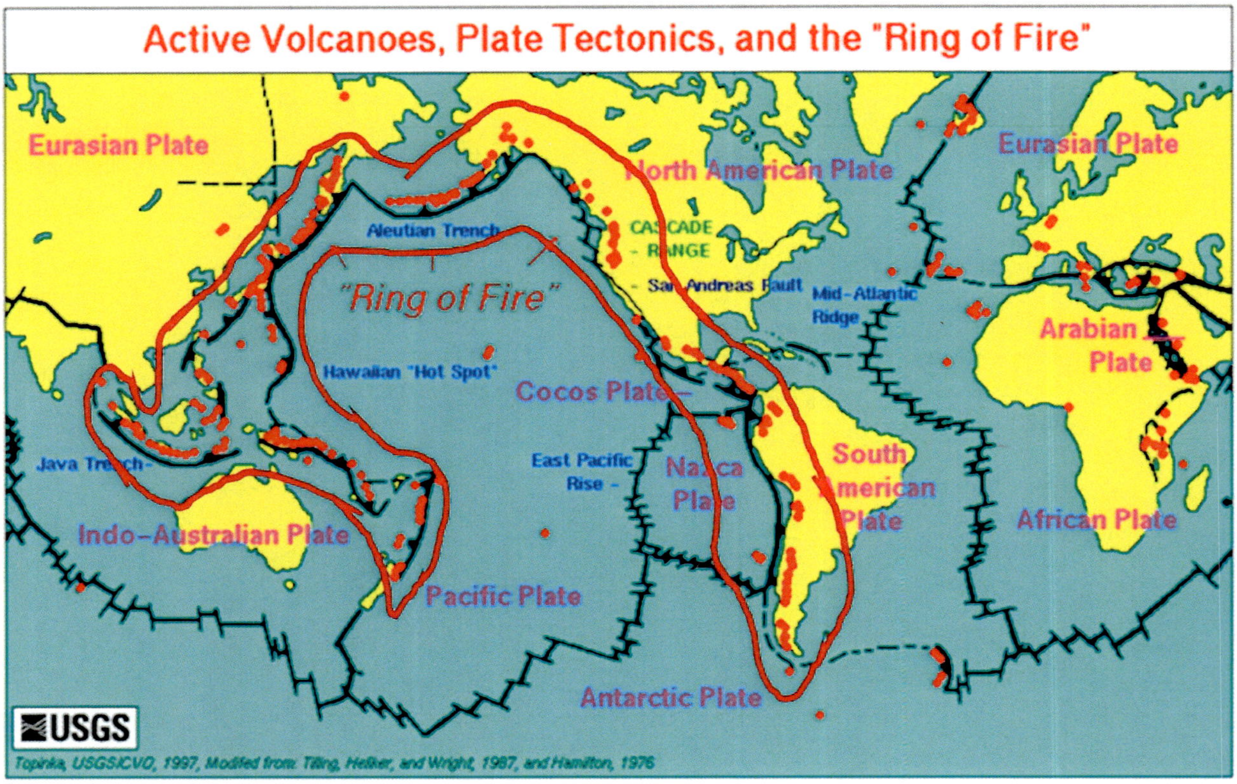

It was only well into the 20th Century that geologists could actually "see" the ocean bottom through special imaging technology. What they discovered amazed them and in an indirect way gave more credence to the Genesis account of the Flood. The ocean floor reveals cracks all around the globe. These huge cracks are what geologists call ***oceanic trenches*** or rifts. Some of these dwarf the Grand Canyon in size and depth. A good explanation for these cracks is that they were part of the breaking of the fountains of the great deep recorded in Genesis 7:11.

In line with many of these cracks in the earth are located the volcanically active chains of earthquakes and volcanoes such as the Ring of Fire. Modern secular geologists have divided these cracks into plates that they claim have been moving apart for millions of years. The idea is called Plate Tectonics. But it is only an idea. It has never been observed or scientifically verified.

A Biblical way to look at this map is simply to observe what seems to be the remnant of what had been the pre-Flood Earth, now scarred and altered by a massive flood catastrophe, the likes of which will not happen again. The earthquakes, volcanoes, climate changes and extremes are simply the vestiges of a violent time in Earth's recent past through which God judged an evil world.

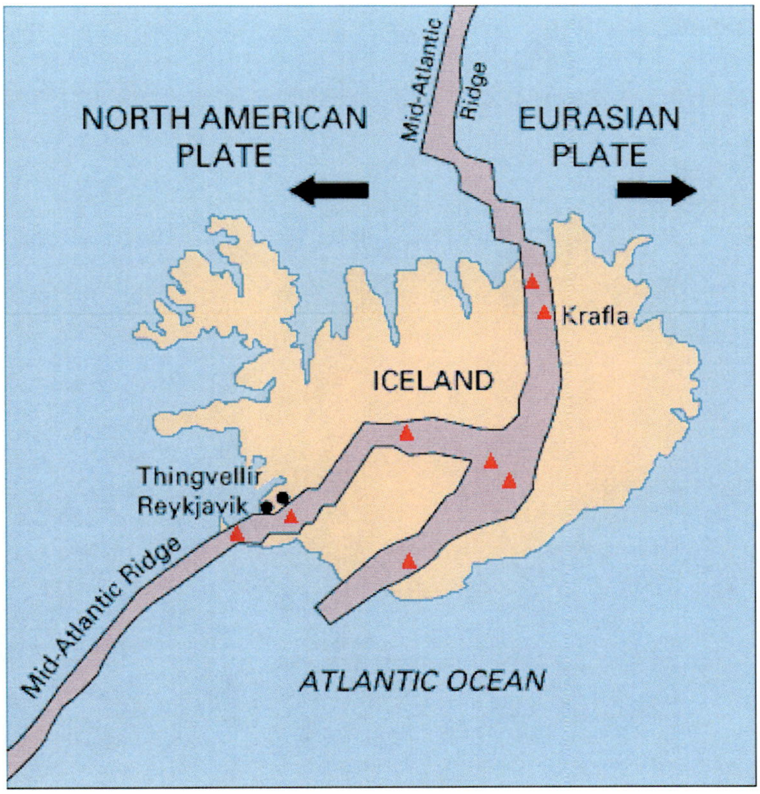

The rift in the center of the map (left) runs right through Iceland.

The Mid-Atlantic Ridge (rift) runs through Iceland, and is a visible surface feature of the ridge. The rift is a popular destination for tourists in Iceland. Some of the rifts are filled with clear water.

Parts of a Volcano

Although volcanoes are different from one to another, they all share some common parts. Study the diagram of a composite volcano below and learn the various parts. Be sure to add unfamiliar terms to your vocabulary section of your notebook. (Some of the definitions are included in the following activity.)

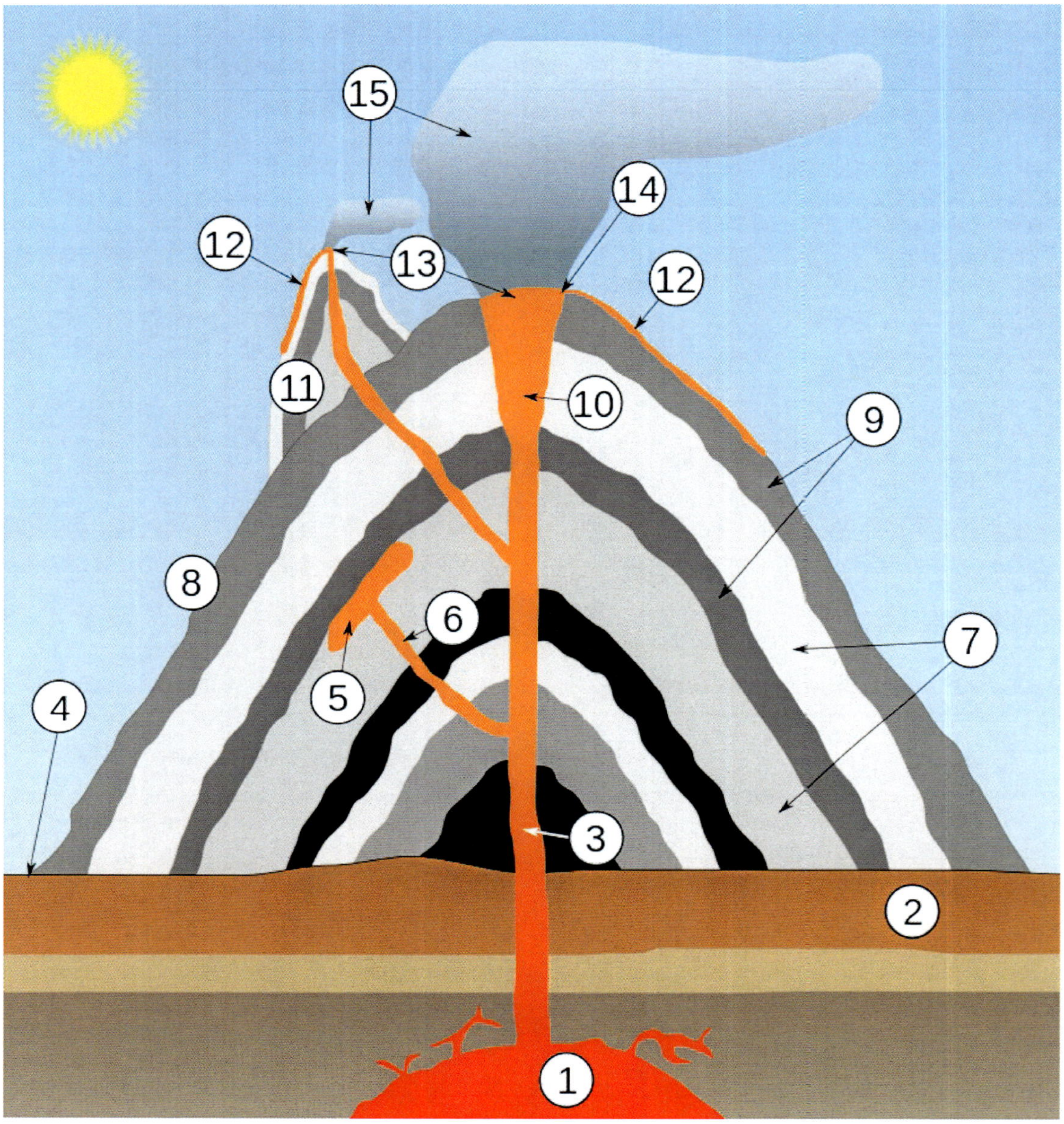

Cross-section through ***a stratovolcano*** *(vertical scale is exaggerated): 1. Large magma chamber 2. Bedrock 3. Conduit (pipe) 4. Base 5. Sill 6. Dike 7. Layers of ash emitted by the volcano 8. Flank 9. Layers of lava emitted by the volcano 10. Throat 11. Parasitic cone 12. Lava flow 13. Vent 14. Crater 15. Ash cloud*

Activity #2
Volcano Diagram

Draw your own volcano diagram, based on the diagram above. Before doing this, study the diagram above, and try to draw your own from memory. You may use any of the terms listed above, but be sure to include the basic parts listed below. Label your diagram.

Ash cloud – the cloud of ash that forms in the air after some volcanic eruptions
Conduit – a passage through which magma (molten rock) flows in a volcano
Crust – Earth's outermost, rocky layer
Lava – molten rock; it usually comes out of erupting volcanoes
Magma chamber – contains magma deep within Earth's crust
Side vent – a vent in the side of a volcano
Vent – an opening in the earth's surface through which volcanic materials erupt

Take Quiz 2

Lesson Three – Types of Volcanic Rocks

Words to know: element mineral rock mafic felsic fine-grained rhyolite andesite dacite phenocryst pyroclastic tephra tuff pahoehoe A'a

Types of Volcanic Rocks

In order to understand and identify volcanic rocks, we will have to learn about what is in them. What makes them black? What makes them gray or pink? Why are some of them filled with holes? Why do some of the volcanic rocks polish to a brilliant shine and others remain dull? Volcanoes don't just erupt black lava, but also a variety of dark and light-colored lavas, some that are very beautiful. All these questions can be answered by studying the minerals that are contained in the volcanic rocks.

The Minerals

Let's start at the beginning and discuss what rocks are made of. ***Rocks*** are made of ***minerals.*** But even smaller yet, minerals are made of ***elements.*** So, we have to talk about which elements are significant to our discussion.

8 Most Abundant Elements in the Earth's Crust

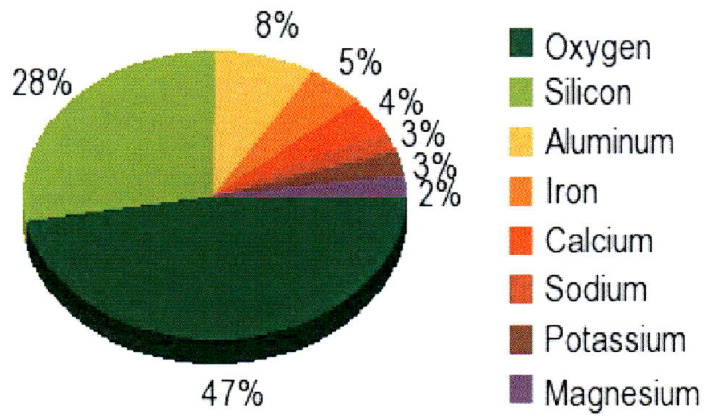

The earth is an amazing place! It is uniquely fitted for life. That shouldn't surprise us. After all, "In the beginning God created the space and the earth." Let's talk about the most abundant elements in God's creation. This chart tells us that out of all the elements (over 90 natural elements), there are eight that form most of the earth's crust and consequently the rocks that comprise the earth's crust. These eight elements make up the minerals. Combinations of these various minerals make up all of the rocks. (Other minerals may be present in the rocks, but these are the primary ones.) On a very interesting note, if you change any one of these elements such that you decrease or increase the percentages, you will change the nature of our earth. God has given us just the right elements in the right percentages to make life possible. You will also notice that these elements are vital for our own health and are found in the foods we eat. *You will need to learn these elements.*

That means that normally you cannot see the mineral makeup with the naked eye, only a general color. The colors are due to the minerals that they contain. So, we will begin with the minerals that make up the volcanic rocks – the rock-forming minerals.

Learning the Chemical Symbols for the Most Abundant Elements

- Oxygen - O
- Silicon - Si
- Aluminum - Al
- Iron - Fe
- Calcium - Ca
- Sodium - Na
- Potassium - K
- Magnesium - Mg

The Rock-Forming Minerals

Even though there are over 4,000 identified minerals today, you only need to know just a few of them in order to understand and identify volcanic rocks.

The Mafic (magnesium and iron) Minerals – the Darker Minerals

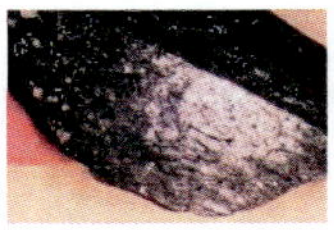 **Amphibole** $(Ca,Na)_{2-3}(Mg,Fe,Al)_5(Al,Si)_8O_{22}(OH,F)_2$	 **Pyroxene** $(Ca,Na)(Mg,Fe,Al,Ti)(Si,Al)_2O_6$	 **Calcium Feldspar** $(Ca,Na)(Si,Al)_4O_8$	 **Biotite Mica** $K(Mg,Fe{+}{+})_3[AlSi_3O_{10}(OH,F)_2$	 **Iron** Fe	 **Olivine** $(Mg,Fe)_2SiO_4$

The Felsic Minerals (feldspar and silica (quartz) – the Lighter Minerals

 **Quartz** SiO_2	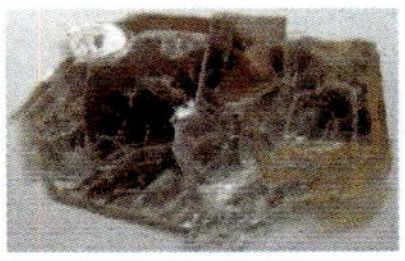 **Muscovite Mica** $KAl_2(Si_3Al)O_{10}(OH,F)_2$	 **Potassium Feldspar** $KAlSi_3O_8$	 **Sodium Feldspar** $NaAlSi_3O_8$	*Sedimentary Mineral*  **Calcite** $CaCO_3$	*Sedimentary Mineral* 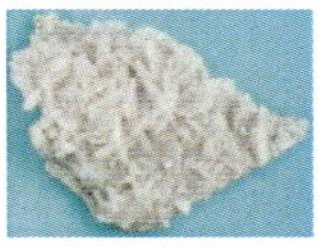 **White Calcite** $CaCO_3$

Notice that all these minerals are made of the eight most abundant elements, combined in a special way to produce specific minerals. We call these minerals, ***the rock-forming minerals*** because they form most of the rocks on Earth. The rock-forming mineral calcite is mainly found in sedimentary rocks. When it is found on volcanic rocks, it is called caliche. It is found primarily on volcanic rocks from arid regions. Why this occurs is not totally understood.

It is these minerals that give lavas their characteristic colors. Here is a rule of thumb to remember when trying to identify volcanic rocks – the darker colored minerals make up the darker colored volcanic rocks and the lighter colored minerals make up the lighter colored volcanic rocks.

The main thing you need to know about identifying volcanic rocks is that they are all ***fine-grained*** rocks. That is, their individual mineral crystals cannot generally be seen with the naked eye, only a general color. From the list below, you will notice that volcanic rocks can be dark colored, very light colored, or in between. These shades of color are due to the rock-forming minerals.

The Darker Colored Volcanic Rocks

The Basalt Family – basalt is a mafic volcanic rock made up of the darker colored minerals from your chart above. It has very little quartz and is quite common. It characterizes the lava flows of Hawaii and of the Columbia Basalt Flows.

a) Aphanitic basalt (plain black or gray lava): it can used as building material. In the pictures below, it was used for a pillar and for building a wall both found at Capernaum, Israel).

b) Vesicular basalt (basalt with gas vesicles*)*

c) Basalt breccia (angular pieces of volcanic rock in a petrified mud matrix; samples from Yellowstone volcanic mudflows)

d) ***Pahoehoe*** basalt (ropy lava, found on the Big Island of Hawaii)

e) ***A'a*** basalt (sharp, spiky lava, found on the Big Island of Hawaii)

f) Scoria (frothy basalt from Hawaii, and El Malpais in New Mexico)

g) Cinders (erupted small particles, like popcorn, from Craters of the Moon in Idaho)

h) Bombs (basalt lava thrown through the air during eruptions, from El Malpais in New Mexico)

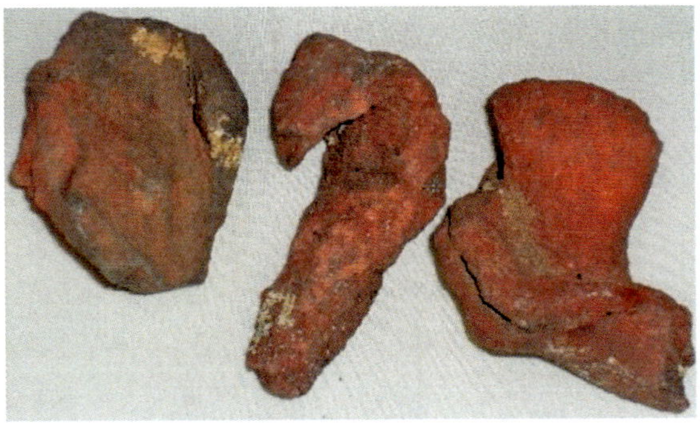

i) Amygdaloidal basalt (vesicular basalt with crystals in the gas pockets found in Oregon)

j) Olivine basalt (the predominant mineral is olivine, which is green, from Hawaii and Arizona)

k) Oxidized basalt (basalt changed by oxygen, from El Malpais and Oregon)

l) Pillow basalt (lava that is erupted underwater, from Washington state)

All these samples of basalt have one thing in common – they are all made of the darker minerals. One of the dark minerals in basalt is iron. Like a rusty nail, basalt changed by oxygen turns orange or rust colored.

The Lighter Colored Volcanic Rocks

The Rhyolite Family – Rhyolite is a felsic volcanic rock made up of the lighter colored minerals from your chart above. It has a lot of both of the minerals, quartz and potassium feldspar. It characterizes the lava from explosive eruptions like Yellowstone National Park.

a) Common rhyolite from Yellowstone.

b) Vesicular rhyolite (the vesicles are filled with silica, from California)

c) Welded ***tuff*** (rhyolite ash and volcanic rock bits welded or fused together because of the tremendous heat involved. The first two samples are from Yellowstone, and the last two are from California.)

d) Pumice (frothy rhyolite glass, from California)

e) Obsidian (rhyolite glassy lava colored by various minerals, like iron. The first two samples are from California, and the others are from Oregon.)

f) Other types of volcanic glass

Perlite (altered rhyolitic glass, often containing nodules of obsidian, from Arizona)

Vitrophyre (glassy lava that has been altered by contact with water, from Yellowstone)

Volcanic nodules (Canada)

Thunder eggs (the state rock of Oregon) are a combination of volcanic glass and rhyolite. The Native Americans believed thunder eggs were the leftover "bombs" that the spirits of the volcanoes had hurled at each other in their anger. Thunder eggs are volcanic in nature, consisting of usually an outer nodule of the extrusive igneous rock, rhyolite, and an inner core of agate, jasper, quartz and or opal. No two thunder eggs are alike. No one knows for sure how they were created, as no one has witnessed a thunder egg being formed under today's geologic conditions. The uniformitarian view states that they were formed over millions of years as various elements in liquid form seeped into the hardened rhyolite structures. But, again, no one has ever seen a thunder egg form. So, the uniformitarian view is not purely a scientific explanation. It is a worldview that attempts to explain the formation of these very interesting rocks.

A more satisfying explanation is to look at things from a Biblical perspective and at conditions on earth before the Genesis flood. As the fountains of the great deep burst open and volcanoes emerged through these great geological upheavals, pockets of molten magma filled with rich mineral solutions were either spewed or were part of the great lava flows which followed. As these globules or gas pockets of molten magma cooled, so did their cores of silica-rich contents, leaving crystallized formations of varying colors and patterns. Thunder eggs are truly unique and wonderful collector items. They are also a testimony to the cataclysmic effects of a world-wide flood.

Thunder eggs from Oregon (Left: exterior matrix; Center and right: interior, cut and polished)

The Intermediate Volcanic Rocks – Volcanic rocks that are in between basalt and rhyolite in color are called intermediate rocks. These include the ***andesite*** family and the ***dacite*** family. They are usually shades of gray or brown, kind of nondescript rocks, and very plain.

The Andesite Family **–** Rocks that contain a mix of the darker and lighter colored minerals and with more quartz than basalt. These rocks characterize stratovolcanic rocks like those that came from Mt. St. Helens. They tend to be gray to brown in color and often show lots of sodium feldspar and hornblende crystals.

The Dacite Family **–** Rocks that contain a mix of the darker and lighter colored minerals, but with more quartz than andesite, are in the dacite family. The dacite family of rocks tend to be gray to pinkish in color and often show flow banding similar to rhyolite. This volcanic rock characterizes violent volcanic eruptions associated with calderas and stratovolcanoes. Sometimes it is hard to distinguish this rock from rhyolite. In fact, some geologists have created another category called *rhyodacite*. Here is a good rule of thumb – if you cannot decide whether your rock is rhyolite or dacite, then very likely it is an intermediate rock.

Porphyritic Volcanic Rocks

All volcanic rocks can contain large visible mineral crystals called ***phenocrysts***. Exactly why this happens is not totally clear to geologists. When volcanic rocks display this feature, we call them, (*volcanic rock name)* ***porphyry,*** as in ***dacite porphyry.***

Basalt porphyry (sodium feldspar crystals), andesite porphyry (sodium feldspar crystals), dacite porphyry (dark hornblende or pyroxene crystals), rhyolite porphyry (potassium feldspar and/or sodium feldspar crystals)

Pyroclastic Volcanic Rocks

The word ***pyroclastic*** means fire broken. The name refers to the broken bits and pieces of rock, ash (glass) and hot steam that roll down the side of a volcano at tremendous speeds, hugging the ground and stripping everything in its path. As these clouds of rock and steam come to rest, they fuse together, often into a hard rock called welded tuff. It is generally a felsic rock, often displaying small bits and pieces of volcanic rock and minerals in its makeup.

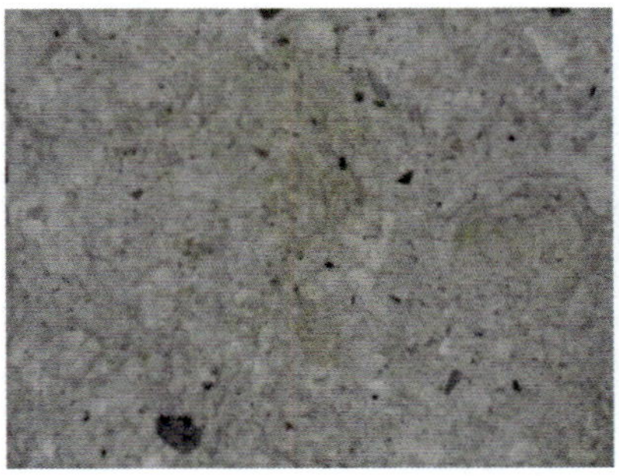

The term pyroclastic is usually reserved for rocks formed from welded ash and bits and pieces of welded rock. But all forms of volcanic ejecta, including tuff, can be called another term, ***tephra***. The following types of ejecta are examples of tephra.

Pele's Tears and Pele's Hair - small pieces of solidified lava drops formed when airborne particles of molten material fuse into tear like drops of volcanic glass. Pele's tears are jet black in color and are often found on one end of a strand of Pele's hair, shown in the second picture.

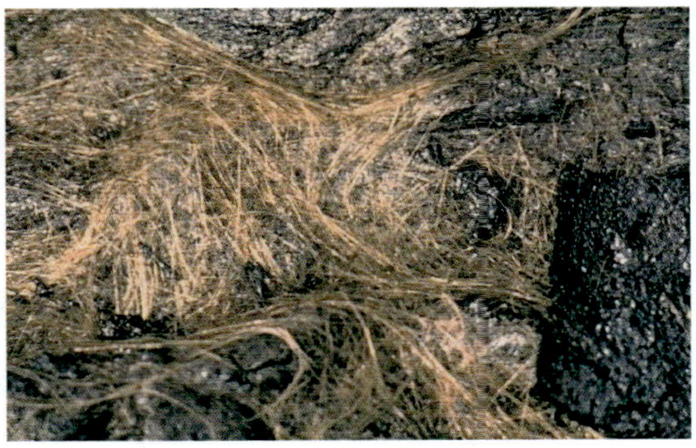

Basalt Bombs - Volcanic bombs are lava fragments larger than 64 mm in diameter that were ejected while still viscous and partially molten.

Ash – Volcanic ash is not like the ash that comes from burning wood or paper. It is a fine cloud of bits and pieces of broken volcanic rock fragments.

Pumice - Pumice is a very light and porous rock, that is the result of super-heated, pressurized, gas-rich rock, that is create3d when violently ejected from a volcano.

Finally, look at the following chart. It will help you to see how the various types of lavas are related to the types of volcanoes that produce them. Notice that the more explosive a volcano is, the more it tends to produce the andesitic and rhyolitic lavas. In the next lesson we will learn more about this tendency as it relates to viscosity of lava.

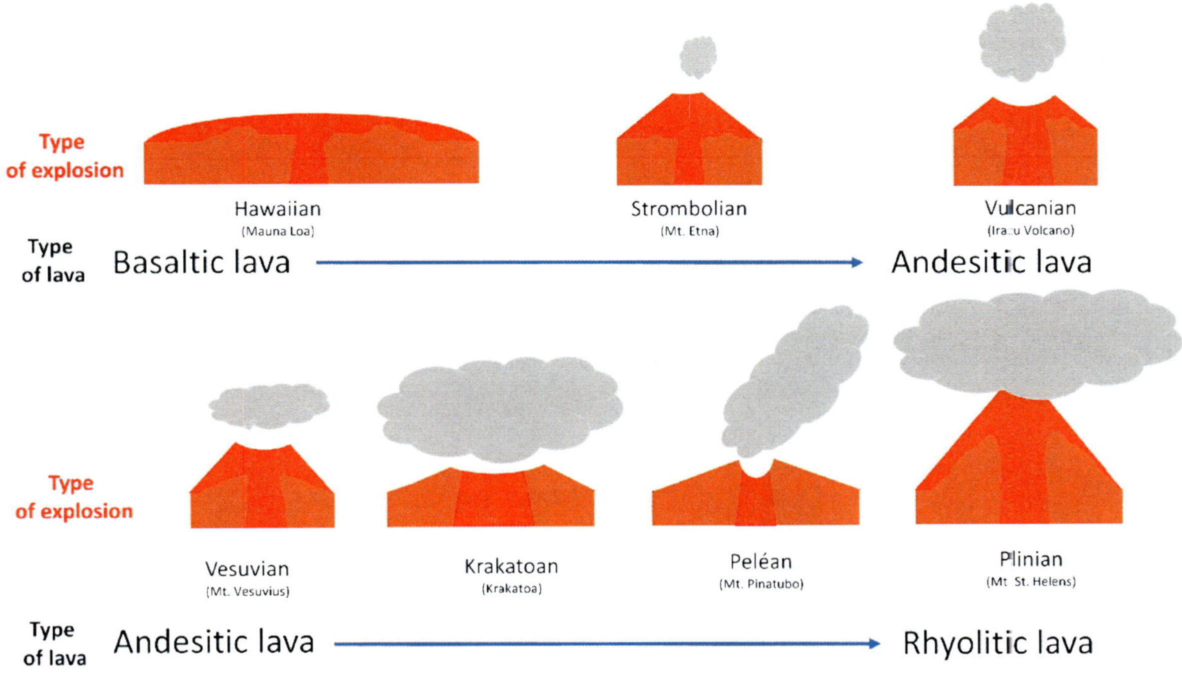

Activity #3
Identifying the Volcanic Rocks

1. Take out the volcanic rocks that came with your kit. Organize them according to light, dark and intermediate rocks.
2. Select one famous volcano and write a report about it. Include what is known about its history, its affects, and the type of rocks it erupted.
3. In the back of your textbook, you will find a page with a variety of volcanic rocks pictured. Cut them out and organize them according to dark colored, light colored and intermediate rocks.

Take Quiz 3

Lesson Four – Types of Volcanic Eruptions

Words to know: phreatomagmatic eruption phreatic eruption basaltic lapilli viscosity volcanology
Volcanic Explosivity Index lava magmatic eruption seamount submarine eruption resistance

Types of Volcanic Eruptions

During a volcanic eruption, ***lava***, tephra (ash, ***lapilli***, volcanic bombs and blocks), and various gases are expelled from a ***volcanic vent*** or fissure. What kinds of factors influence the type of volcanic eruption which will occur? Volcanologists have identified three different ***factors***:

1. *Superheated steam and gas* released under pressure producing ***magmatic eruptions*** characteristic of Hawaiian volcanoes; more on the order of ***basaltic*** showers rather than explosive eruptions like Mt. St. Helens.
2. *Thermal contraction from chilling on contact with water* or from steam explosions causing ***phreatomagmatic eruptions.*** A phreatomagmatic eruption is the result of an explosive *water-magma interaction* characteristic of the eruption of Mt Usu in Japan in April 2000. Mount Usu formed on the southern rim of the caldera containing the lake which came into contact with the hot magma. Large amounts of steam and magmatic gases are emitted.
3. The ejection of volcanic rock from past eruptions that have *blocked the conduit or pipe* of a volcano and have thus been *under great amounts of pressure* causing ***phreatic eruptions.*** A phreatic eruption, also called a phreatic explosion or ultravulcanian eruption, occurs when magma heats ground or surface water. In fact, its name means, "of or pertaining to water." The extreme temperature of the magma (anywhere from 932 to 2,138 °F causes near-instantaneous evaporation to steam, resulting in an explosion of steam, water, ash, rock, and volcanic bombs. At Mount St. Helens, hundreds of steam explosions preceded the 1980 explosive eruption.

All three of these factors involve the ***interaction of magma and water*** to some degree or another. Is it any wonder that during the great Genesis Flood many violent volcanoes would have been active and catastrophic such as the Yellowstone Caldera?

Types of Eruptions

Types of eruptions have been classified and named after historically significant volcanic eruptions.

- ***Hawaiian eruptions -*** Fluid basaltic lava is erupted into the air as spray from a vent or fissure. These eruptions can last for hours or days. The lava that is erupted produces cinders, scoria, and lava flows. These tend to be the calmest of the volcanic eruptions. The cones that are built tend to form as shields rather than as high cones.

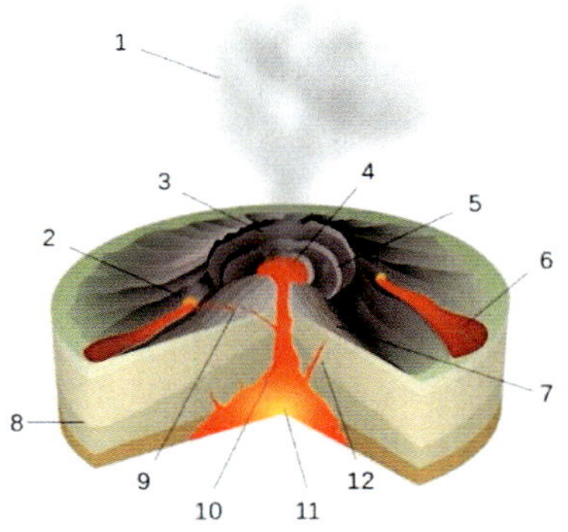

Hawaiian Eruption and Pahoehoe Lava from Kilauea
Diagram of a Hawaiian eruption key: 1. Ash plume 2. Lava fountain 3. Crater 4. Lava lake 5. Fumaroles 6. Lava flow 7. Layers of lava and ash 8. Stratum 9. Sill 10. Magma conduit 11. Magma chamber 12. Dike.

- ***Strombolian eruptions - named after the volcano Stromboli, which has been erupting continuously for centuries.*** These types of eruptions are characterized by high gas content and episodic blasts of lapilli and volcanic bombs. They form mostly viscous, basaltic lava. They are similar to the Hawaiian eruptions, but noisier. Strombolian eruptions are often associated with small lava lakes, which can build up in the conduits of volcanoes. They are one of the least violent of the explosive eruptions, although they can still be very dangerous if bombs or lava flows reach inhabited areas.[3] Some of the products produced by these eruptions are fluid lavas, scoria, Pele's tears, and Pele's hair.

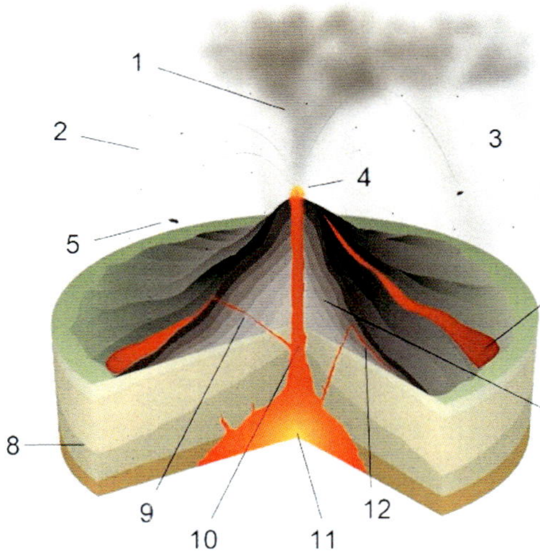

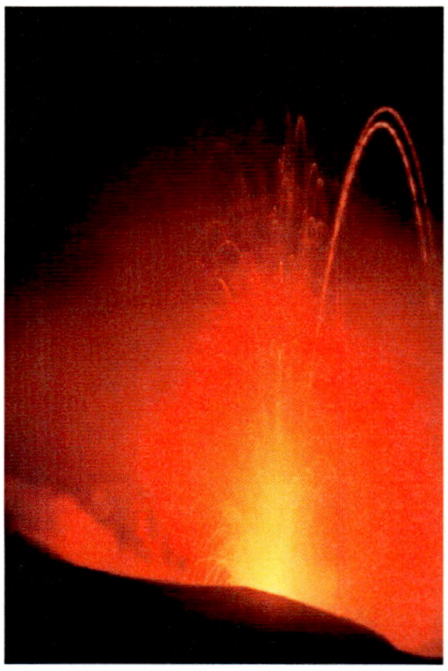

Diagram of a Strombolian Eruption and Mt. Stromboli, Italy
Diagram of a Strombolian eruption key: 1. Ash plume 2. Lapilli 3. Volcanic ash rain 4. Lava fountain 5. Volcanic bomb 6. Lava flow 7. Layers of lava and ash 8. Stratum 9. Dike 10. Magma conduit 11. Magma chamber 12. Sill

[3] https://geology.com/volcanoes/types-of-volcanic-eruptions/.

- ***Vulcanian eruptions –*** named for the Italian island of Vulcano. A Vulcanian eruption is a short, violent, relatively small explosion of viscous magma (usually andesite, ***dacite***, or ***rhyolite***). This type of eruption results from the fragmentation and explosion of a plug of lava in a volcanic conduit, or from the rupture of a lava dome. Vulcanian eruptions create powerful explosions in which material can travel faster than 800 mph and rise several miles into the air, producing tephra, ash clouds, and pyroclastic flows. Vulcanian eruptions may be repetitive and go on for days, months, or years, or they may precede even larger explosive eruptions.[3]

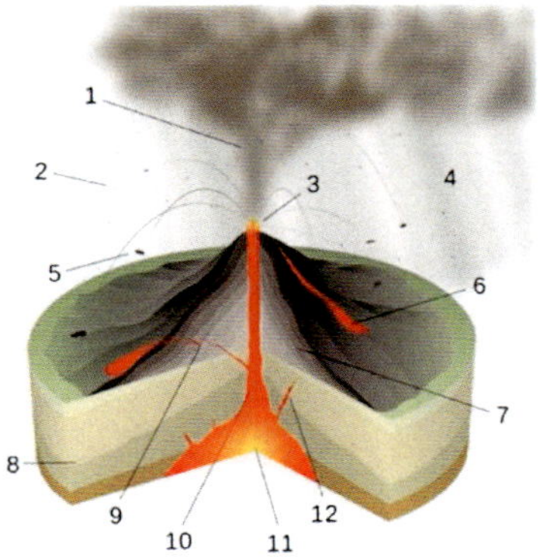

Diagram of Vulcanian Eruption and Tavurvur in Papua New Guinea in the process of erupting
Diagram of a Vulcanian eruption key: 1. Ash plume 2. Lapilli 3. Lava fountain 4. Volcanic ash rain 5. Volcanic bomb 6. Lava flow 7. Layers of lava and ash 8. Stratum 9. Sill 10. Magma conduit 11. Magma chamber 12. Dike

- ***Peléan eruptions –*** named after the volcano Mount Pelée in Martinique, the site of a massive Peléan eruption in 1902 that was one of the worst natural disasters in history. A Pelean eruption is associated with explosive outbursts that generate pyroclastic flows, dense mixtures of hot volcanic fragments and gas. The flows produced by these eruptions are heavier than air but are of low ***viscosity*** and travel at great velocities. As a result, they are extremely destructive.[4]

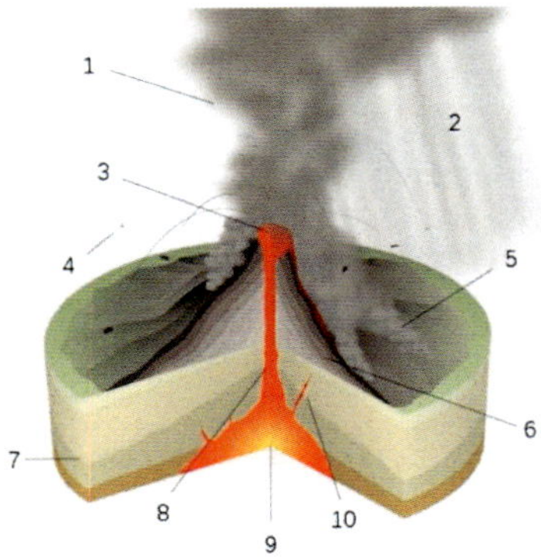

Diagram of a Pelean Eruption and Mayon Volcano, Philippines, 1984
Diagram of Peléan eruption key: 1. Ash plume 2. Volcanic ash rain 3. Lava dome 4. Volcanic bomb 5. Pyroclastic flow 6. Layers of lava and ash 7. Stratum 8. Magma conduit 9. Magma chamber 10. Dike

[4] https://www.britannica.com/science/volcano/Six-types-of-eruptions

- ***Plinian eruptions (or Vesuvian)*** – named for the historical AD 79 eruption of Mount Vesuvius that buried the Roman towns of Pompeii and Herculaneum. These eruptions are the largest and most violent of all the types of volcanic eruptions. The eruptions are caused by volatile gases which are stored in the magma, and are usually associated with very viscous magmas (dacite and rhyolite). They release enormous amounts of energy and create eruption columns of gas and ash that can rise up to 35 miles high, travelling at speeds of hundreds of feet per second. Ash from an eruption column can drift or be blown thousands of miles away from the volcano. The eruption columns are usually shaped like a mushroom (similar to a nuclear explosion) or an Italian pine tree; Pliny the Younger, a Roman historian, made the comparison while viewing the 79 AD eruption of ***Mount Vesuvius***, and Plinian eruptions are named for him. Plinian eruptions are extremely destructive, and can even obliterate the entire top of a mountain, as occurred at ***Mount St. Helens*** in 1980. They can produce falls of ash, scoria and lava bombs miles from the volcano, and pyroclastic flows that will sandblast whatever is in their path.[5]

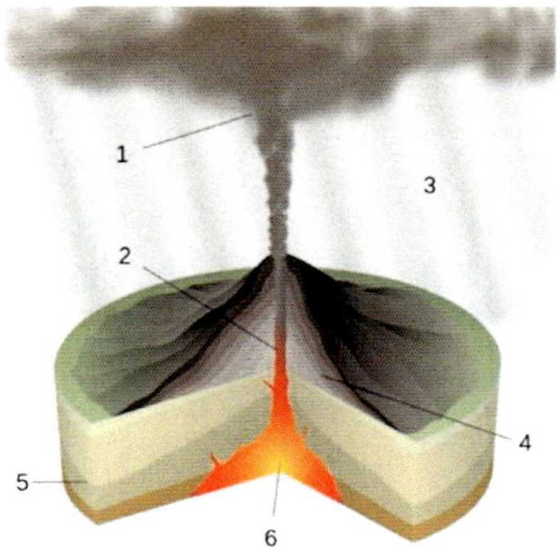

Diagram of a Plinian Eruption and the April 21st, 1990 eruptive column from Redoubt Volcano in Alaska, as viewed to the west from the Kenai Peninsula. Diagram of a Plinian eruption key: 1. Ash plume 2. Magma conduit 3. Volcanic ash rain 4. Layers of lava and ash 5. Stratum 6. Magma chamber

- ***Surtseyan eruption (also called hydrovolcanic or phreatomagmatic eruptions*** – a kind of hydromagmatic eruption, where magma or lava interacts explosively with water. In most cases, Surtseyan eruptions occur when an undersea volcano has finally grown large enough to break the water's surface; because water expands when it turns to steam, water that comes into contact with hot lava explodes and creates plumes of ash, steam and scoria. Lavas created by a Surtseyan eruption tend to be basalt, since most oceanic volcanoes are basaltic. The volcanic island of Surtsey is a classic example of a Surtseyan eruption. It erupted between 1963 and 1965 off the coast of southern Iceland. Hydromagmatic activity built up several square kilometers of tephra over the first several months of the eruption; eventually, seawater could no longer reach the vent, and the eruption transitioned to Hawaiian and Strombolian styles.[6]

---

[5] https://geology.com/volcanoes/types-of-volcanic-eruptions/

[6] https://geology.com/volcanoes/types-of-volcanic-eruptions/

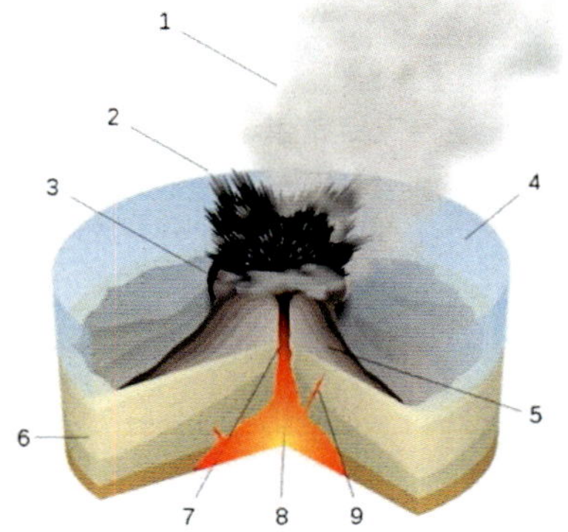

Diagram of a Surtseyan Eruption and the Island of Surtsey, erupting 13 days after breaching the water. A tuff ring surrounds the vent. Diagram of a Surtseyan eruption key: 1. Water vapor cloud 2. Compressed ash 3. Crater 4. Water 5. Layers of lava and ash 6. Stratum 7. Magma conduit 8. Magma chamber 9. Dike

- ***Submarine eruptions – are underwater fissures from which magma erupts.*** An estimated 75% of the total volcanic of submarine eruptive volume is generated by submarine eruptions near mid ocean ridges alone. Because of the problems associated with detecting deep sea volcanoes, they remained virtually unknown until advances in the 1990s made it possible to observe them. Although most are located in the depths of seas and oceans, some also exist in shallow water, which can spew material into the air during an eruption. The lava formed by submarine volcanoes is quite different from terrestrial lava. Upon contact with water, a solid crust forms around the lava. Advancing lava flows into this crust, forming what is known as pillow lava.[7]

Geologists think that submarine eruptions may have produced ***seamounts*** or table mounts. These are extremely interesting in that they are essentially flat-topped underwater volcanoes. Now the question – how did they become flat-topped? The receding stage of the Flood could account for the flat tops of these volcanoes with its shearing force of the water. They indicate that these were at one time above water. Do you suppose they may have something to do with the Genesis Flood?

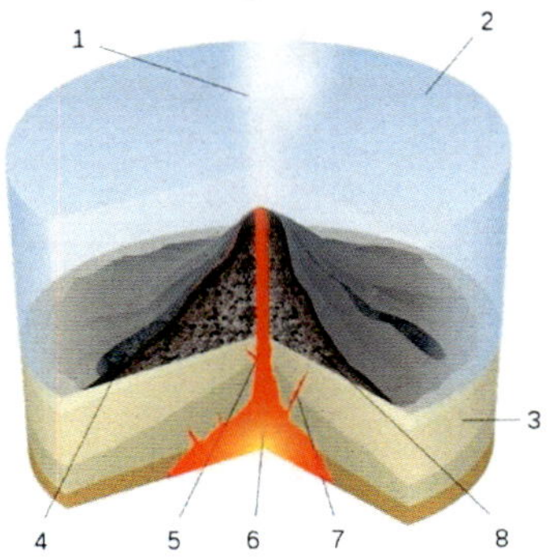

Diagram of a Submarine Eruption and Lō'ihi, a seamount, or underwater volcano, on the flank of Mauna Loa, the earth's Largest Shield Volcano Diagram of a submarine eruption key: 1. Water vapor cloud 2. Water 3. Stratum 4. Lava flow 5. Magma conduit 6. Magma chamber 7. Dike 8. Pillow lava

---

[7] http://dictionary.sensagent.com/Submarine_volcano/en-en/

- ***Phreatic eruptions*** - are steam-blast eruptions. They area steam-based volcanic reactions caused by the heating of water by magma. The pressure caused by underground steam forces it to escape through volcanic eruption. The water sources of phreatic eruptions include both underground and surface water. The volcanic explosion is caused by the extremely rapid vaporization of water into steam, which occurs when underground magma is in the close vicinity of the water source. These temperatures can reach up to 1,700 degrees Celsius. Phreatic explosions typically cause the dispersal of rock and water material. While it is common for many types of eruptions to cause the dispersal of magma, it is unusual in phreatic eruptions. Phreatic eruptions that disperse magma are known as phreatomagmatic eruptions.[8]

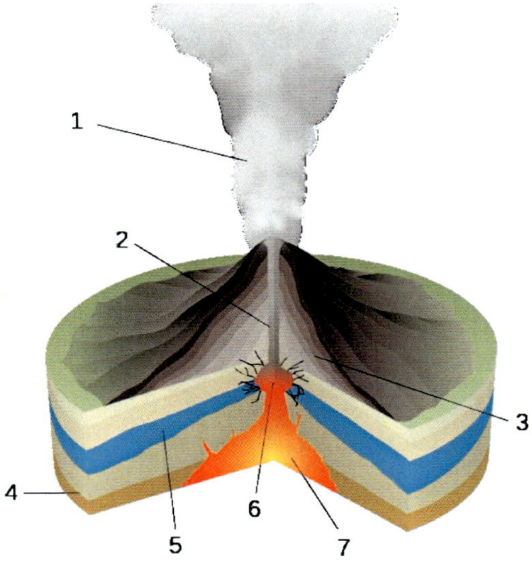

Diagram of a Phreatic Eruption and the Eruption of Mt. St. Helens, 1980
Diagram of a phreatic eruption key: 1. Water vapor cloud 2. Magma conduit 3. Layers of lava and ash 4. Stratum 5. Water table 6. Explosion 7. Magma chamber

[8] https://www.reference.com/science/phreatic-eruption-fc3afb1f543ede4b

The following diagram is a summary comparison of some of the various types of explosions.

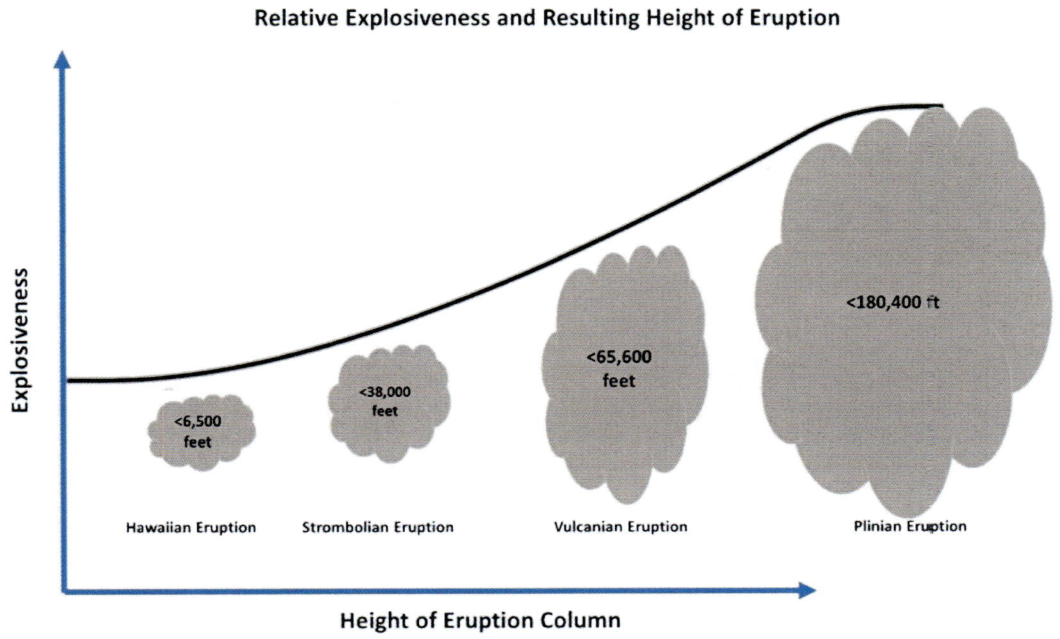

Measuring Volcanic Eruptions

Volcanology is a special branch of geology that studies volcanoes. In order to monitor, record and evaluate volcanic eruptions, volcanologists have come up with an ***Volcanic Explosivity Index (VEI)*** for measuring the power of volcanic explosions. This index is based on types of historical eruptions

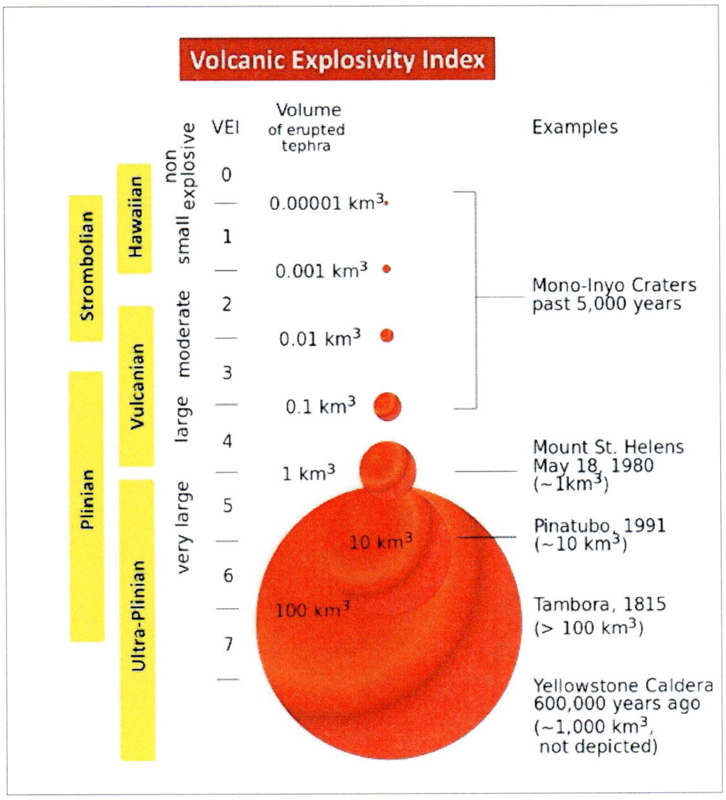

- Hawaiian – the most mild
- Strombollian
- Vulcanian
- Plinian
- Ultra-Plinian – the most explosive

Viscosity

Another way of measuring a volcano's potential for destruction is to measure the flow rate of ***lava***. This is especially important for evacuating populations that may be in the vicinity of a volcanic flow. A modern example of this occurred on the Hawaiian Island of Hawaii. In 1986 a new vent opened from Kupaianaha fissure and eventually covered the town of Kalapana in January 1991. The flow could be predicted because volcanologists could measure its ***viscosity***. Viscosity is a measurement of the ***resistance*** to flow that lava possesses. The mineral *quartz* determines how viscous lava is. Lavas with more quartz tend to be highly viscous. In other words, they are pasty and move rather slowly. Another factor is the temperature of the lava. Hotter lavas are less viscous and move more quickly.

Different fluids have different viscosities ((low to high viscosity) honey, oil, lava).

The temperature of lavas with high viscosity (i.e., rhyolite) is much less than lavas with low viscosity, (i.e., basalt*)*. Basalt is the type of lava that covered the town of Kalapana, Hawaii. Below is a map showing some of the historical lava flows around Kalapana, (July 1985-February 1992). The town was eventually over run by the lava.

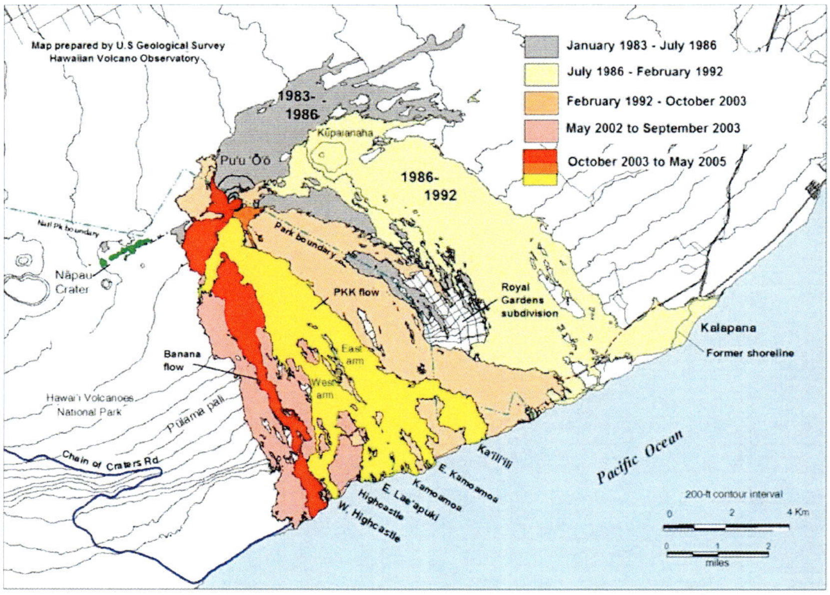

Kalpana buried by lava flow, 1990.

Volcanologists have developed a scale for measuring viscosity and gas content from the various types of eruptions. Quartz consistently seems to be related to the viscosity of lavas. The more quartz, the more explosive an eruption. Gas content can provide volcanologists with information as to pyroclastic flows from an eruption. The abundance of gases varies considerably from volcano to volcano. Water vapor is consistently the most common volcanic gas, normally comprising more than 60% of total emissions. Carbon dioxide typically accounts for 10-40% of emissions. Although basalt often appears in vesicular form due to gas content, the amount of quartz is much lower than in andesite or rhyolite and hence the basaltic lava is not as explosive.The following chart is a comparison of gas content and viscosity by volcano type.

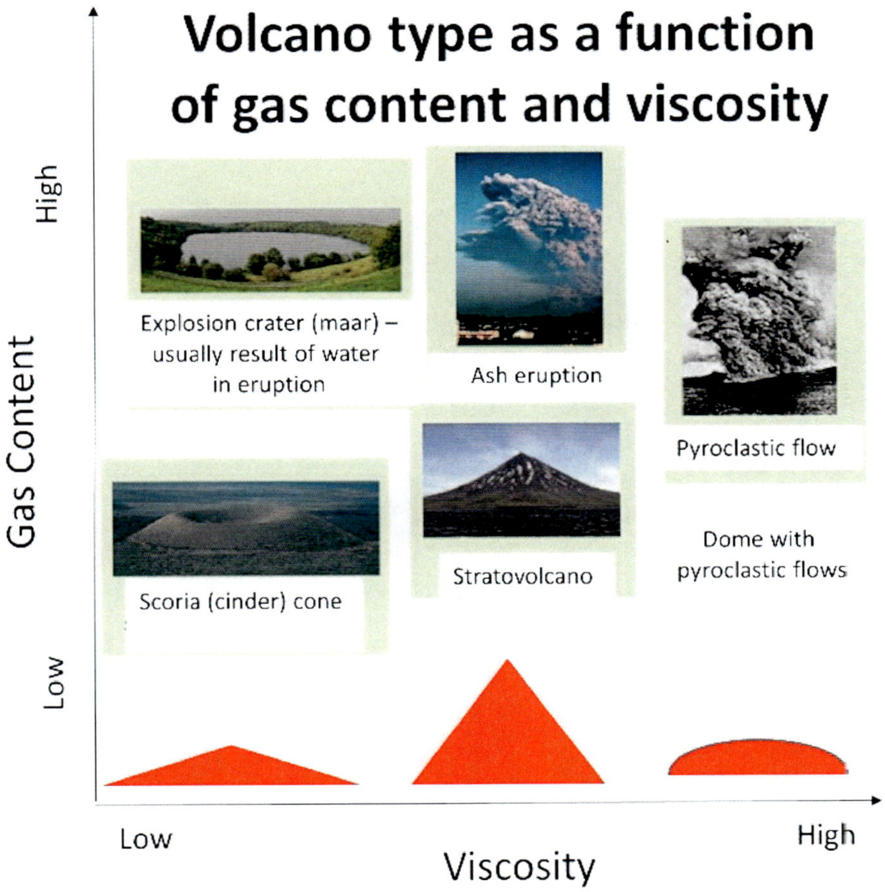

Combining various indexes volcanologists have developed will give us a chart like this one below. Because geologists know the mineral content of each type of lava, they can determine the viscosity of lava and consequently the potential for violent explosions. This kind of information is useful in protecting human life and property. But it does not tell us anything about the origin of volcanoes nor of their remote history.

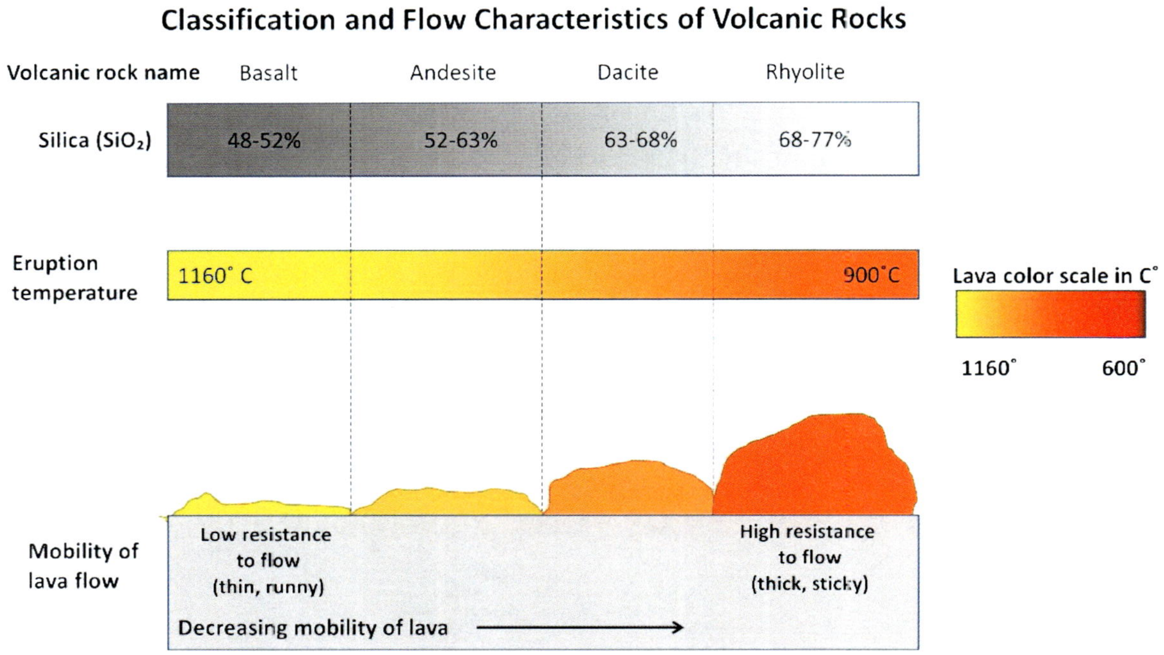

From the chart above, what is the primary factor in determining lava's viscosity? It is the amount of the mineral quartz. Quartz has the chemical formula, SiO_2 – silicon dioxide.

Activity #4(a)
Drawing the Volcanic Eruption Types

Draw and color your own pictures illustrating the different volcanic eruption types. Label the various parts of the eruption.

Activity #4(b)
Charting the properties of viscosity

Instructions: For this project you will need

- Several sheets of various kinds of paper, like wax paper, copy paper, construction paper, etc.
- Cookie sheet
- Graph paper for recording your results
- Several different kinds of fluids, like ketchup, mustard, honey, molasses, pop, etc. (You can work with as many different fluids as you like. The purpose is to take note of how different fluids flow and what factors influence the viscosity of a fluid)

Procedure:

- Give each fluid a letter or number for identification.
- Place the paper you will use on your project on the cookie sheet, Place one teaspoonful of each fluid onto the piece of paper. Tilt it at an angle so that the fluids will move. For more variety, tilt the paper at various angels. Be sure to be consistent with each fluid you experiment with. Record what you observe. Did the fluid flow slowly or quickly? How far did it flow? Now try the same procedure with a different kind of paper and then with different angles. Record what you observe.
- For variation you can try refrigerating the fluids and heating the fluids and repeat the above process for each variable. Are there differences in flow rate? Does temperature affect the viscosity of the fluid?
- Summing it all up: What kinds of factors influenced the viscosity of the different fluids?

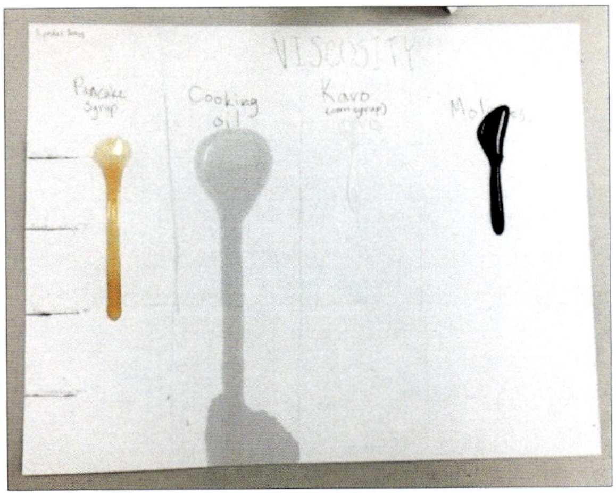

Take Quiz 4

Lesson Five – Major Volcanic Eruptive Events in History

Words to know: trap (volcanic)   volcanic arc   oceanic plateau   volcanic province   plateau

Volcanic eruptive events have occurred in history on a scale that is not occurring today. And in fact, violent eruptions have been steadily declining. Even secular geologists understand this, as the following chart illustrates. (Note: The scientific evidence supports the sizes and intensities of past volcanic eruptions, not the ages. The chart is only used for comparison of past volcanic activity. How geologists arrive at the long ages given for these volcanoes will be discussed in ***Lesson Six, Radiometric Dating and Volcanic Rocks.)***

The following illustration shows the change throughout history relative to amount of ejecta and size of eruption. Ignoring the secular dates given for these eruptions, the main point is that through time they have been steadily decreasing in intensity.

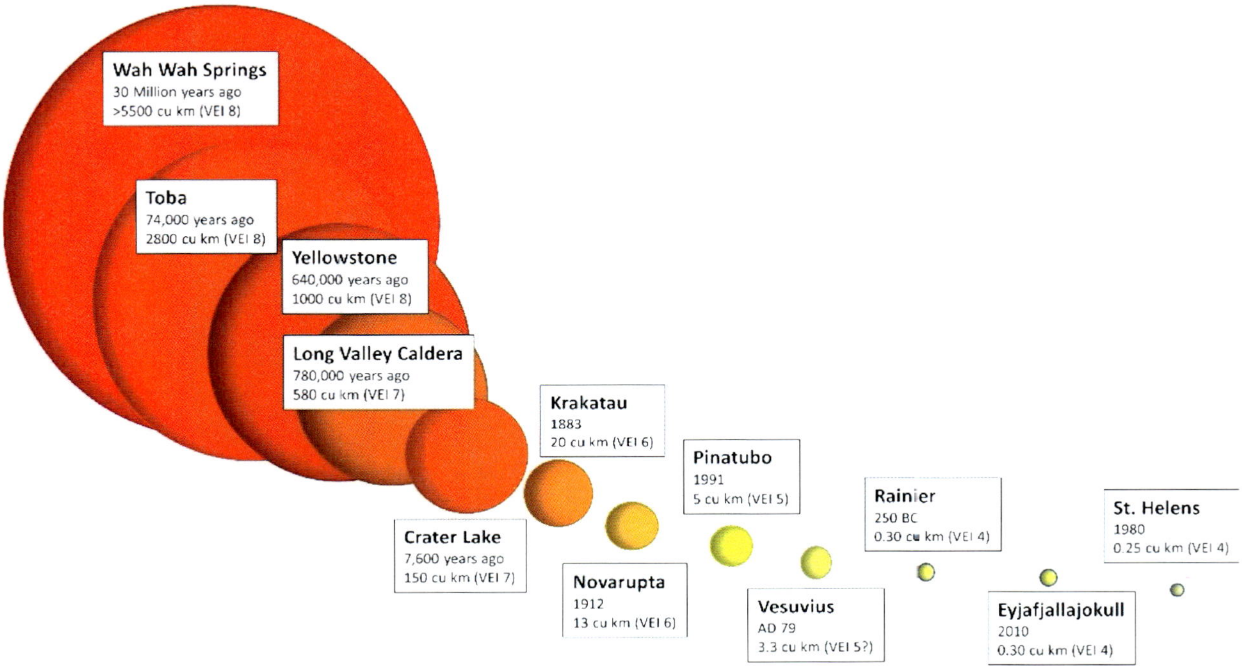

Flood and post-Flood volcanic eruptions were significantly greater in the past than what we experience today. Although volcanic activity has subsided, we are still technically in the post-Flood period. Volcanoes still erupt but are now a faint reminder of what took place during and shortly after the Flood.

Let's take a look at some of the major eruptive events in history.
Nd

1. ***The Columbia Flood Basalts*** **-** The Columbia River Basalt Group is a large ***volcanic province*** that lies across parts of the Western United States. It is found in the U.S. states of Washington, Oregon, Idaho, and Nevada and covers over (81,000) 63,000 square miles. Some of the lava in this region is over a mile thick. There are no obvious volcanic cones in this area so naturally the question arises as the where this lava came from. Geologists today are not certain where the lava came from, but conjecture that fissures opened at some

point and extruded the lava. Genesis 7:11 could certainly explain how this might have happened.

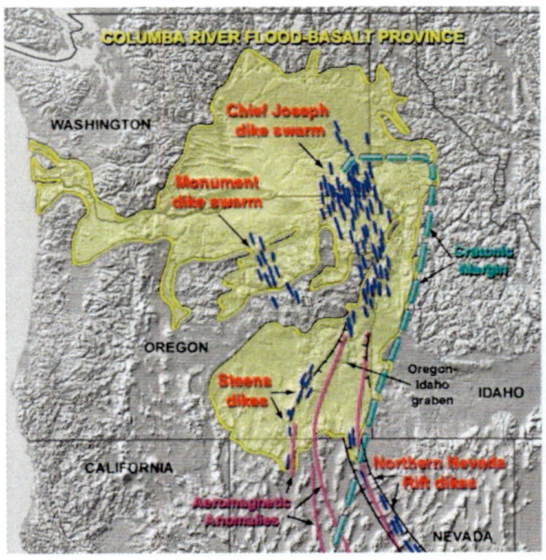

2. Similar to the Columbia Flood Basalts are the ***Deccan Traps***. The term ***trap,*** used in geology for such rock formations, is derived from the Dutch word for stairs and refers to the step-like hills forming the landscape of the region. They consist of multiple layers of solidified flood basalt that together are more than 6,000 feet thick and cover an area of around 200,000 square miles. That is three times the size of the Columbia Flood Basalt formation in the United States!

In fact, there are several such basalt flows around the world. Check out the map below. The areas marked in blue are huge basalt flows that have gone on in the past. This kind of activity is not happening today.

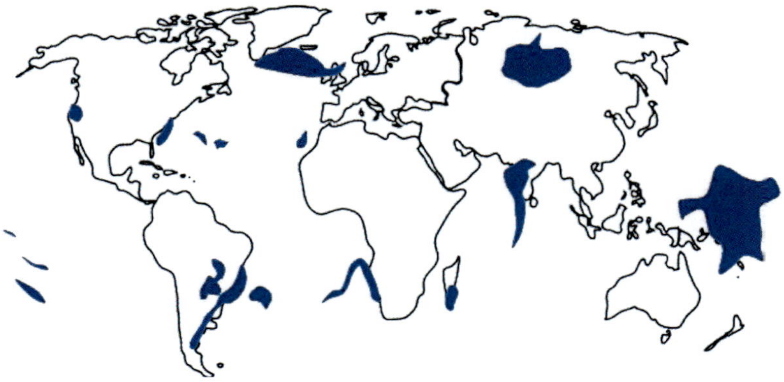

3. ***The Yellowstone Supervolcano*** – considered to be the largest volcanic eruptions known in history. Three eruptions probably back to back produced calderas as large as 72 miles across! Such massive volcanic eruptions would have produced huge amounts of ash, tuff, and lava flows. They covered vast miles of land and are a few miles thick in some places. Catastrophic eruptions like these are not occurring today.

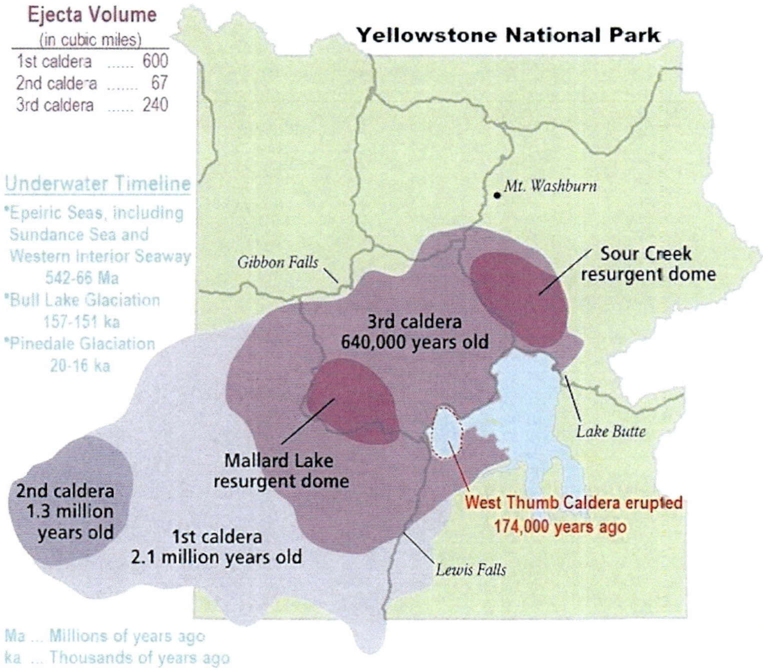

ADAPTED WITH PERMISSION FROM *WINDOWS INTO THE EARTH* BY ROBERT SMITH AND LEE J. SIEGEL, 2000.

Henry Heasler and Cheryl Jaworowski, US NPS, 2013.

Although the ages for these eruptions on this illustration are not in line with a Biblical view, the size of the eruption would fit with the Biblical account of a global catastrophic flood. We will cover more on the age of the caldera when we talk about radiometric dating.

4. ***The Absaroka Volcanic Range of Mountains*** – 150 miles long, this range of weathered volcanic rock testifies to the awesome power of volcanic eruptions that would have taken place during the last stages of or shortly after the global flood of Genesis. All these examples provide us with clues as to how an ice age that covered much of the earth shortly after the flood, could have been triggered. (We will talk more about this is ***Lesson Ten***.)

Pilot and Index Peaks in the Absaroka Mountains

5. ***La Garita Caldera*** – a large caldera in southwestern Colorado. The is one of the largest explosions in recent geologic history. The resulting deposit, known as the Fish Canyon Tuff, has a volume of approximately 1,200 cubic miles.[9] By comparison, the May 18, 1980 eruption of Mt. St. Helens was only 0.25 cubic miles in volume.

Ash and tuff beds, La Garita

6. ***Mt. Mazama (Crater Lake, Oregon)*** – Mount Mazama is a destroyed volcano in the Oregon part of the Cascade ***Volcanic Arc*** (chain of volcanoes), part of the Ring of Fire, and part of the Cascade Range located in the United States. The volcano's collapsed caldera holds Crater Lake, and the entire mountain is located within Crater Lake National Park. Mazama was destroyed by a volcanic eruption. The eruption created a caldera that is about a mile deep. Much of the volcano fell into the volcano's partially emptied neck and magma chamber.[10]

[9] Mason, Ben G.; Pyle, David M.; Oppenheimer, Clive (2004). "The size and frequency of the largest explosive eruptions on Earth". Bulletin of Volcanology. ***66*** (8): 735–748. Bibcode:2004BVol...66..735M. doi:10.1007/s00445-004-0355-9.
[10] https://pubs.usgs.gov/fs/2002/fs092-02/

7. ***Mt. Tambora, Dutch East Indies, 1815*** – one of the largest volcanic eruptions in recorded history. Its eruption alone caused global temperatures to fall a couple of degrees. The year 1816 has been called the year without a summer as many countries experienced crop failure due to frost or lack of sunshine.[11]

This image was taken by the NASA Expedition 20 crew.

8. ***The Giant's Causeway*** **-** on the north-east coast of Northern Ireland is a great testament to the awesome amount of basalt that flowed during the Genesis Flood. There are about 40,000 hexagonal basalt pillars that have formed here.[12] This basalt lava flow extends for many miles.

[11] https://scied.ucar.edu/shortcontent/mount-tambora-and-year-without-summer

[12] Information found at: https://www.ireland.com/en-us/amazing-places/giants-causeway/?utm_source=bing&utm_medium=cpc&utm_campaign=TI_US_EN_PRO_EM_All%20Island_NA_IOI_NA_NA_NA_ACP_Icons-Bing&utm_term=giants%20causeway%20map&utm_content=Icons%7CGiant%27s%20Causeway&gclid=COaP-Y7i_-ACFUaSxQIdWUoFhw&gclsrc=ds

9. ***Tamu Massif*** - Resting on the bottom of the Pacific Ocean sits Tamu Massif, the largest volcano in the world at 400 miles wide and 2½ miles tall. It has created a new class of volcano called an ***oceanic plateau*** – a plateau built of massive lava flows. It has been estimated that it is around 120,000 sq. miles.[13] The entire volcano is bigger than the British Isles or New Mexico. Geologists now recognize that this size of volcanic eruption would have produced outpourings of lava thought to have been linked to mass extinctions and climate change. Adding this volcano to the list above makes it pretty clear that the Genesis Flood was more than simply a local flood. It was a huge event that triggered such subsequent events as Tamu Massif.

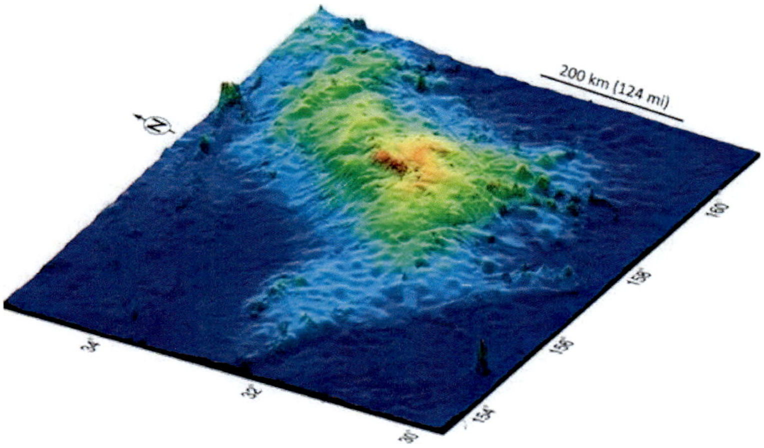

These are just a few of the tremendous catastrophic volcanic activities of the past. There is nothing like them occurring today! Secular geologists tell us that, "The present is the key to the past." But there is nothing like these volcanic events happening today that we can use as a model for interpreting the past history of the earth. In spite of these obvious examples, secular geologists continue to believe that, "The present is the key to the past." These volcanic activities must have been associated with a unique historical geologic event that triggered them. The Genesis Flood offers just such a trigger, and is a good explanation for what we see today.

---

[13] *Tamu Massif: Largest Volcano on Earth Discovered Beneath Pacific Ocean*, Enrico de Lazaro, September 9, 2013; found at http://www.sci-news.com/othersciences/geophysics/science-tamu-massif-largest-volcano-pacific-01366.html.

Activity #5
Charting the Major Eruptive Events

Using Mt. St. Helens as a recent volcanic eruption, construct a chart showing the comparison of the amount of volcanic material extruded from the various volcanic events in history with that of Mt. St. Helens.

Take Quiz 5

Lesson Six – Radiometric Dating and Volcanic Rocks

Words to know: radiometric dating   decay   geologic time table   nucleus   neutron   proton   electron   radioactivity   isotope   metamorphic   theoretical   periodic table   assumption   bias   leaching   relative dating   absolute dating   extrapolate

What is Radiometric Dating?

Many geologists believe that they can tell how old a rock is by a process known as radiometric dating. Radiometric dating is an idea that involves measuring the present rate of the radioactive ***decay*** of certain radioactive elements. This present observed rate of decay is then ***extrapolated*** into the past to arrive at age of a rock. There are two basic types of radiometric dating. Carbon-14 dating measures the decay of carbon-14 from things that were once alive, like charcoal or paper. The other type of dating involves rocks that geologists think were once molten, like volcanic rocks or in some cases, ***metamorphic*** rocks. In both cases it is the present observable rate of decay that is assumed to have been constant throughout the history of the item being dated. Now, here is the question: Can geologists really measure the rate of radioactive decay in a volcanic rock such that they can tell how old a rock is?

How Radiometric Dating Works

First, let's get a basic idea of how radiometric dating is supposed to work. To understand radiometric dating, you need to understand atoms and their makeup. All atoms are made of particles called ***electrons, protons***, and ***neutrons***. These atoms form the elements of the ***periodic table***. Most of these are stable elements that God has given us for life. They can all be nicely arranged according to specific properties.

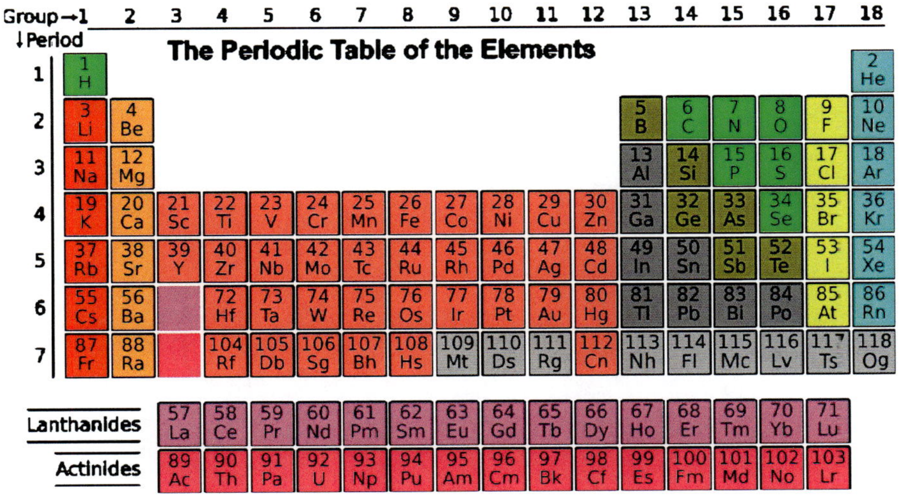

Normal atoms are finely tuned organizations of these atomic particles. ***Radioactivity*** is essentially an imbalance of these atomic particles.

An unstable version of the original element is called an ***isotope.*** The word isotope means *same place* because it occupies the same place on the periodic table as its stable counterpart. As examples of a radioactive isotope is carbon ("C") 14. It has two more neutrons in its ***nucleus*** and so is different than the normal carbon 12 seen in the periodic table. The isotopes are radioactive, some more dangerous than others. This means that the isotopes have a tendency to "lose" particles and this is called radioactive decay. When this happens, energy is released, and this release of energy can be observed and measured. Radiometric dating is based on this decay of a radioactive element. Take a look at the following illustration for an idea of what happens to radioactive elements. When a radioactive element loses particles, it ***theoretically*** can become something else. I say theoretically because some decay rates require amounts of time that have never been observed by humans. It is simply a matter of mathematics and statistics.

Let's take a normal carbon element. It is fairly small and so will make a good example to study. The image on the left represents a normal carbon element. Normal carbon has six protons, six electrons and six neutrons, carbon-12, as it is referred to. It is stable.

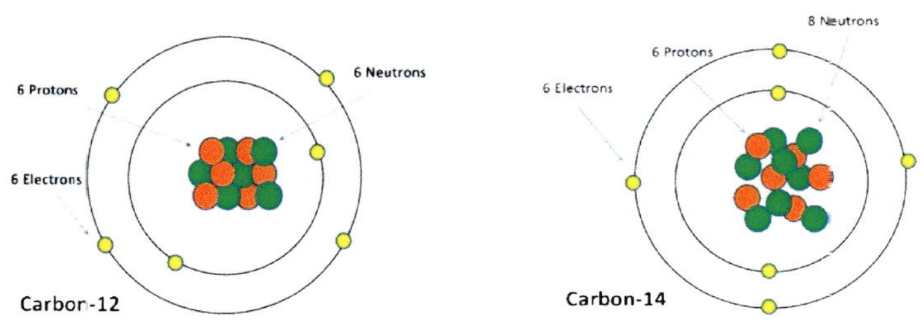

The atom pictured next to it is called carbon-14 and it is abnormal. Why? In this case it has too many neutrons: eight. Physicists have not really been able to discover either why there are unstable elements or why they tend to decay. But they do.

I personally think that normal atoms are the stuff of the created universe and that God made them that way. But then the Fall took place in Genesis 3 and ever since then nothing works as it is supposed to. We live in a dying, decaying world looking for God's redemption and a new heavens and a new earth. Statistically carbon-14 decays or sheds the extra neutrons and changes into something else – stable nitrogen-14. Again, I say statistically because no one has been around long enough to actually measure and prove this process. But by taking away certain atomic particles, it can be shown on paper that atoms behave in this way. This is called radioactive decay.

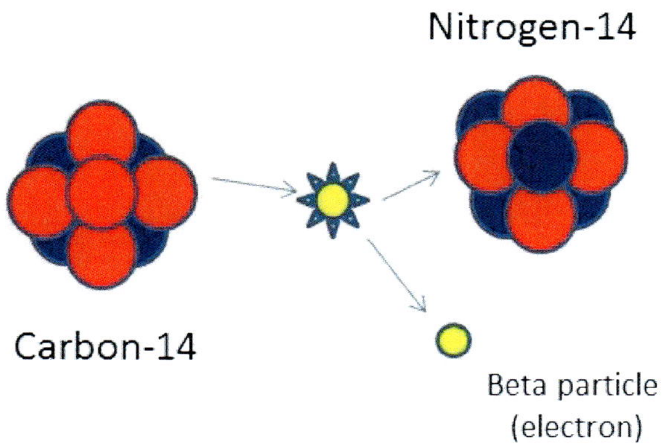

Radiometric dating works on the ***assumption*** that this decay rate has remained constant throughout the remote past. The principle works exactly the same for unstable potassium 40 to stable argon 40 which is what is used to date volcanic rocks in an attempt to decipher their age.

So here is the question: If radioactive decay is a scientific observation, why is it challenged as a method for determining the age of the earth?

There are two issues. The first issue concerns the Scriptures. They are recorded history, and verifiable history. They tell us that the earth is young. ***To date, no discovery has ever overturned the historical reliability of the Scriptures. They have only been rejected out of a bias against them.*** If the Scriptures are what they state that they are (God's revelation to man about a 6,000-year-old earth), then there has to be a problem somewhere in the radiometric dating system.

The second issue with radiometric dating concerns its reliability. There have been an increasing number of situations that reveal inconsistent and just plain wrong dates. Why, if radioactivity is an observed phenomenon, would dates be unreliable and inconsistent? The short answer is that the radiometric dating system relies on assumptions *that can't be proven*. Here is a list of some pretty big assumptions that geologists make when using radiometric dating methods:

1. The original amount of unstable element is *assumed* to have remained the same since its inception, except for its decay.
2. The stable element, known as the *daughter* element, is *assumed* to have all come from the decay of the *parent*, or unstable, element.
3. The decay rate is *assumed* to have remained the same throughout the life of the process.
4. The purity of the system has been *assumed* to have remained the same throughout the process – no ***leaching*** out of the parent element nor transportation in of daughter element.
5. The Biblical position of a historic global flood is rejected as a legitimate explanation of the geology of our earth and therefore *assumed* to be just a myth.

So the big question concerning these assumptions is how would one ever go about verifying that these assumptions are correct? No one has been around long enough to witness the process from beginning to end, be it 50,000 years, 500,000 years or 4.6 billion years, unless you accept the Biblical account of history. This account does have historical eye witnesses and seems to explain a great number of geological phenomena.

In the end, if the date arrived at from a radiometric dating method is older or younger than what is the expected outcome, then someone or something else must be the arbiter as to which radiometric date is correct. That arbiter is called the geologic time table and it has been the "Bible" of geology since the 1800's. It determines what age should be assigned to the rock.

The Geologic Time Table (Scale or Column)

Take a look at the following chart. You have probably seen this in a number of books. It is presented in our schools as ***fact*** and taught as ***truth***. But is it? A little bit of history will help us clear a few things up.

In the 1800s as this chart was first being put together, there were no ages, dates, periods or eras assigned to it. It looked very much like the first column below. The different names like Cambrian and Devonian were regions where certain geologists of the time studied the rocks. It was discovered that these regions contained fossils of creatures – some apparently extinct and some still living. The question was how to arrange the various groups of rocks being studied into a coherent whole. The table became stacked based on three ideas:

1. Life must be far older than what the Bible states
2. Since some fossils represent creatures no longer living, life must have developed from simple to complex over great ages
3. Since strata on the bottom represent layers that were laid down before those on top of them, they must be older – far older than what the Bible states

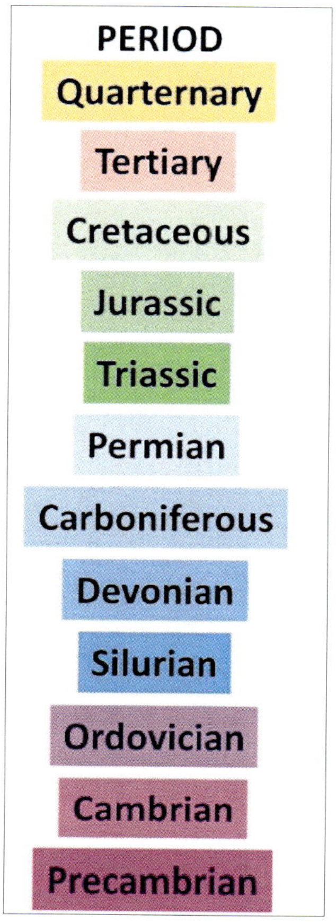

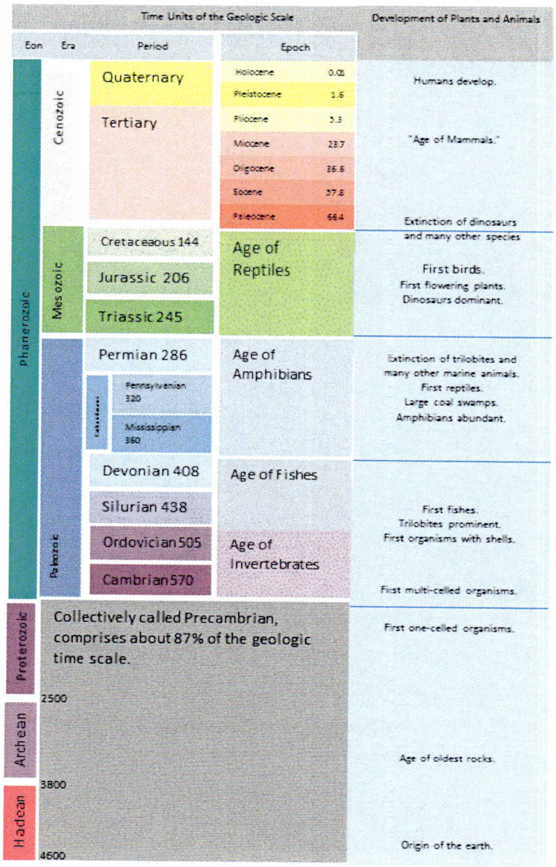

The geologic time scale began as a way to organize fossils being discovered. Obviously, there are no dates or ages written on any fossils. Ultimately, the Scale was organized as a *time* column based on *assumptions* and beliefs about the past history of the earth.

This scale was set long before radioactivity was even discovered. So, there was *no objective, scientific* way to validate the geologic time scale except by way of assumptions. These assumptions, of course, cannot be scientifically tested. In short, a philosophy about life determined how to arrange the fossils and how old they were. This scale is the "Bible" of modern geology. It determines where rocks and fossils should be placed in the history of the earth. It determines how old things should be. If there are apparent contradictions with it, the scale is used like a law. It decides what the answers should be. Radiometric dating came after the development of this scale. And it is the geologic time scale that decides if the radiometric dates are in harmony with what is expected from the geologic time scale.

In geology there are two types of dating of rocks. There is ***relative dating*** and ***absolute dating***. Simply stated relative dating is knowing which rocks were laid down first. This principle goes all the way back to Nicolas Steno (the "father of stratigraphy") in the late 1600s. Steno stated that unless otherwise disturbed, the rocks on the bottom were laid down first. That is really just common sense. And it does not have anything to do with the actual age of the rocks, only the order in which they were laid down. By the 1800s secular geologists were using this principle to infer long ages for Earth history. Let's take a look at a practical example from something known as the Great Unconformity near Yellowstone National Park.

In the picture to the left, volcanic rocks lie over limestone. Steno's principle would state that the limestone was laid down first. But how old is this formation? That must depend on the worldview you start with, for there are no dates imprinted on the rocks. If like Hutton you are convinced that the earth is much older than what the Bible states, then you will ascribe millions of years to this formation. If on the other hand you are convinced that the Biblical record of a global flood took place, then you would consider this formation to be no older than 4,500 years.

Absolute dating would ascribe radiometric dates to the formation based on the radioactive decay within lavas. Absolute dating seeks to give specific numbers to the age of these rocks. But how could you ever be confident or even verify that the radioactive process continued for millions of years with no change?

Are there examples of volcanic rocks for which we know the age, but that have been dated too young or too old? Yes, there are a number of examples.

Radiometric dating gives many surprises. Basalts from Hualalai in Hawaii, observed to have erupted in 1800–01, gave potassium-argon (K-Ar) ages ranging from 160 million years to 3,300 million years. A lava dome on Mt St Helens in USA, observed to form since the 1980 eruption, gave K-Ar ages between 350,000 and 2,800,000 years. Lava erupted from Mt Ngauruhoe, New Zealand, between 1949 and 1975, gave K-Ar ages up to 3,500,000 years.[14]

From just these examples alone, we can see that something is not quite right with the dependability of radiometric dating. Radioactivity is a fact. Radiometric dating is not. The assumptions that geologists have relied on have been shown to be unreliable. It is like depending on a watch that ticks but does not accurately keep time.

To learn more about different rate discrepancies, see ***Appendix B***.

[14] Tas Walker, "A Giant Cause" This article is from *Creation*_27(2):28–34, March 2005. Found at https://creation.com/a-giant-cause

Activity #6
Understanding Radiometric Dating

1. Look at the following chart of the periodic table of elements. Select five stable elements from the table and, using outside resources, find five matching unstable elements. We already studied one – carbon-12 and carbon-14. Draw a picture illustrating the differences between the stable and unstable elements you chose.

The Periodic Table of the Elements

Group→ / ↓Period	1	2	3	4	5	6	7	8	9	10	11	12	13	14	15	16	17	18
1	1 H																	2 He
2	3 Li	4 Be											5 B	6 C	7 N	8 O	9 F	10 Ne
3	11 Na	12 Mg											13 Al	14 Si	15 P	16 S	17 Cl	18 Ar
4	19 K	20 Ca	21 Sc	22 Ti	23 V	24 Cr	25 Mn	26 Fe	27 Co	28 Ni	29 Cu	30 Zn	31 Ga	32 Ge	33 As	34 Se	35 Br	36 Kr
5	37 Rb	38 Sr	39 Y	40 Zr	41 Nb	42 Mo	43 Tc	44 Ru	45 Rh	46 Pd	47 Ag	48 Cd	49 In	50 Sn	51 Sb	52 Te	53 I	54 Xe
6	55 Cs	56 Ba		72 Hf	73 Ta	74 W	75 Re	76 Os	77 Ir	78 Pt	79 Au	80 Hg	81 Tl	82 Pb	83 Bi	84 Po	85 At	86 Rn
7	87 Fr	88 Ra		104 Rf	105 Db	106 Sg	107 Bh	108 Hs	109 Mt	110 Ds	111 Rg	112 Cn	113 Nh	114 Fl	115 Mc	116 Lv	117 Ts	118 Og

Lanthanides	57 La	58 Ce	59 Pr	60 Nd	61 Pm	62 Sm	63 Eu	64 Gd	65 Tb	66 Dy	67 Ho	68 Er	69 Tm	70 Yb	71 Lu
Actinides	89 Ac	90 Th	91 Pa	92 U	93 Np	94 Pu	95 Am	96 Cm	97 Bk	98 Cf	99 Es	100 Fm	101 Md	102 No	103 Lr

2. Look up the meaning of *isotope*. What does it mean?
3. Using an internet resource, go to www.creation.com and find three articles on radiometric dating. Record the radiometric dating methods being written about and also the problems with them.

Take Quiz 6

Lesson Seven – Volcanic Hazards and Emissions

Words to know: lahar (debris avalanche) acid rain ash fall pyroclastic flows noxious gas sulfur dioxide volcanic hazard carbon dioxide hydrogen sulfide volcanic gas (volcanic) emission

Mt. St. Helens

Before the eruption of May 18, 1980, geologists noticed increased activity at Mt. St. Helens in Washington state. On March 20, an earthquake of a magnitude of 4.2 earthquake shook the area, followed by steam venting on March 27. By the end of April, a noticeable bulge could be seen on the north side of the mountain. On May 18, with little warning, a second earthquake, of magnitude 5.1, triggered a massive collapse of the north face of the mountain. Prompting the largest known ***debris avalanche*** in recorded history. The Volcanic Explosivity Index for this eruption was rated at a 5 (a Plinian eruption). Some of the results of this explosion were:

- The pyroclastic flow destroyed vegetation and buildings over 230 square miles.
- More than 1.5 million metric tons of ***sulfur dioxide*** was released into the atmosphere.
- The collapse of the northern flank of Mt. St. Helens mixed with ice, snow, and water to create ***lahars*** (volcanic mudflows). Flowing for miles down the Toutle and Cowlitz Rivers, many bridges and lumber camps were destroyed.
- 3.3 billion cubic yards of material was carried into the Columbia River by the mudflows, 17 miles to the south.
- The ash plume reached 12 to 16 miles above sea level and moved eastward at an average speed of 60 miles per hour with ash reaching Idaho by noon. By the next morning, ash had reached parts of Alberta, Canada.
- Spokane, Washington, 250 miles away, experienced total darkness from the ash.

- The ash cloud spread across the US in three days.
- The eruption released 24 megatons of thermal energy and 0.67 cubic miles of material.
- The explosion ejected more than 0.67 cubic miles of material.
- The height of the mountain was reduced by about 1,300 feet, leaving a crater 1-2 miles deep and ½ mile wide.
- Over 200 homes, 185 miles of highway and 15 miles of railway.
- 57 people were killed, along with 700 big game animals and (estimated) 12 million fish from a hatchery.
- Between 1980-1986, a new lava dome began forming.
- Numerous eruptions of ash and steam have continued since the 1980 eruption.[15]

The picture below shows just how powerful the eruption of Mt. St. Helens was – enough to blow about 2,000 feet of rock and debris off of its summit!

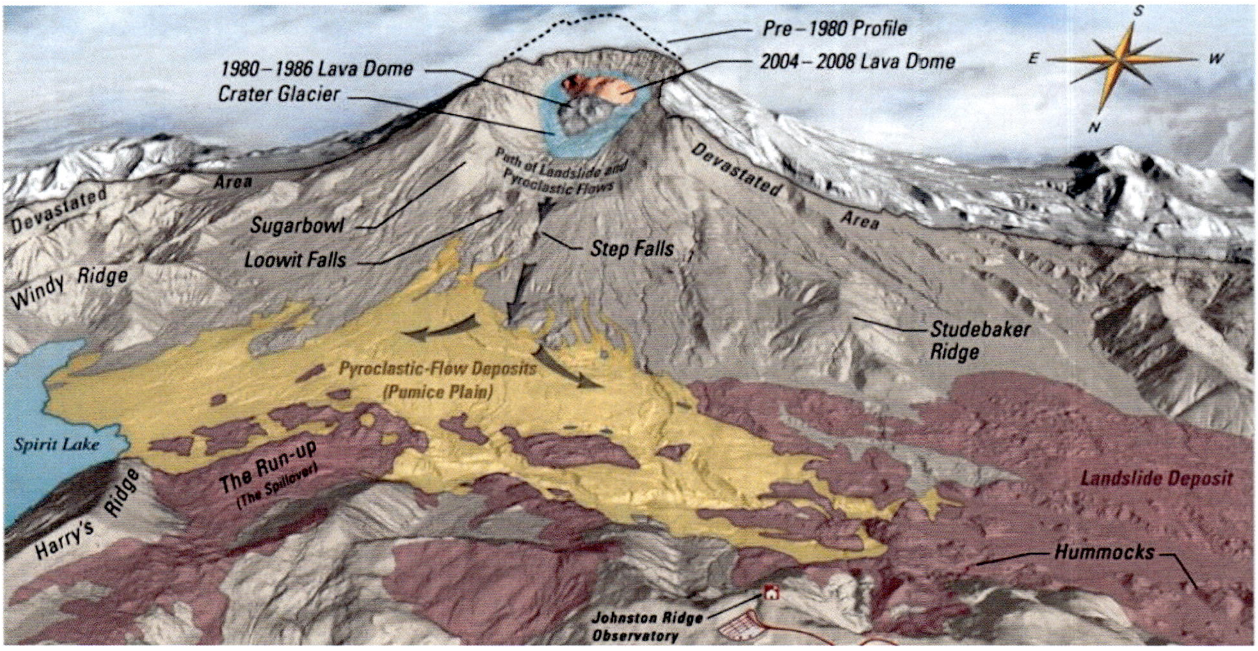

The story of Mt. St. Helens serves is a very instructive one. The volcano has been studied more in the 20th century than any other volcanic eruption. The eruption caused a tremendous amount of damage and shows us that the earth is wearing out.

Secular geologists have told us that the volcanic activity like that of Mt. St. Helens has been going on for billions of years, shaping and molding the earth for life. How could this be given the hazards produced by volcanoes? And secular geologists tell us that Mt. St. Helens is but a small version of what has been going on for billions of years. Let's take a look at the hazards of volcanoes and try to put them in the right framework.

[15] Information was compiled from the USDA Forest Service, and the USGS web sites.

Volcanic Hazards

According to the United States Geological Survey, here are the main things we need to be worried about in a volcanic eruption:

a) ***Acid rain*** – acidic water from the steam that is part of a volcanic eruption. The acid is primarily sulfuric acid, a very caustic and destructive acid. Take a look at the pH scale for measuring acids and bases. Volcanic aerosols contain a good deal of caustic stuff. Notice the destructive caustic effect of the acid rain on the statue.

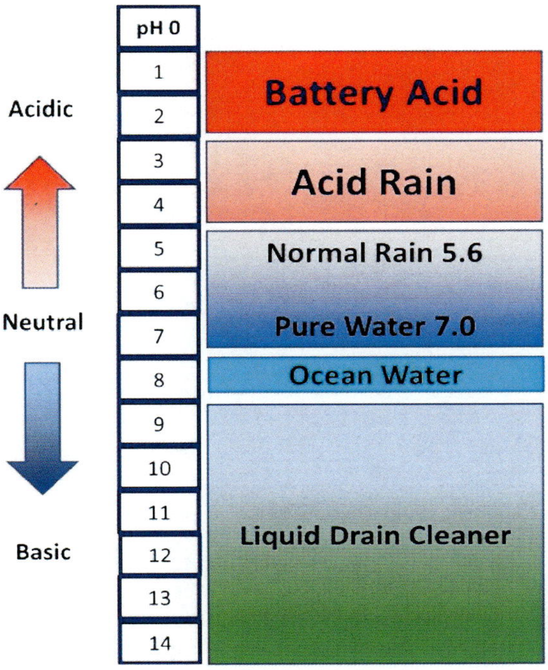

pH scale for acids and bases; Statue destroyed by acid rain

b) ***Debris avalancnes*** or ***Lahars*** (volcanic rock and debris landslides) – these alone have been responsible for destroying towns and villages.

The aftermath of a lahar from the 1982 eruption of Galunggung, Indonesia; Debris avalanches or Lahars (volcanic rock and debris landslides) – these alone have been responsible for destroying towns and villages.

c) ***Ash fall*** – this is not like the ash that comes from burning wood or paper. This stuff is a mixture of bits and pieces of old volcanic rock and glass. Ash destroys machinery and causes lung disease in animals that breathe it and eat the plants on which the ash falls. The ash clouds can travel in the atmosphere for days or weeks, circling the globe many times.

Teleoceras in volcanic ash, Ashfall Fossil Beds, northeast Nebraska; these fossil mammals show signs of a disease in their bones due to the ingestion of volcanic ash, probably as a result of eating ash-covered plants.

d) Pyroclastic flows **–** these are very hot gas clouds filled with glass, steam and bits of older volcanic rock that hug the ground, moving along at speeds of up to several hundreds of miles an hour and stripping everything in their paths.

Mayon Volcano, Philippines, 1984: Pyroclastic flow

The trees in the photo below were all broken like tooth picks and laid down like "pick-up" sticks as a result of the pyroclastic flow of Mt. St. Helens.

e) ***Lava flows*** – we have already seen what these can do to towns.

f) ***Noxious gasses – carbon dioxide*** and sulfur dioxide are deadly gasses. They not only choke living animals, but when mixed with water produce acids, as the pictures below indicate.

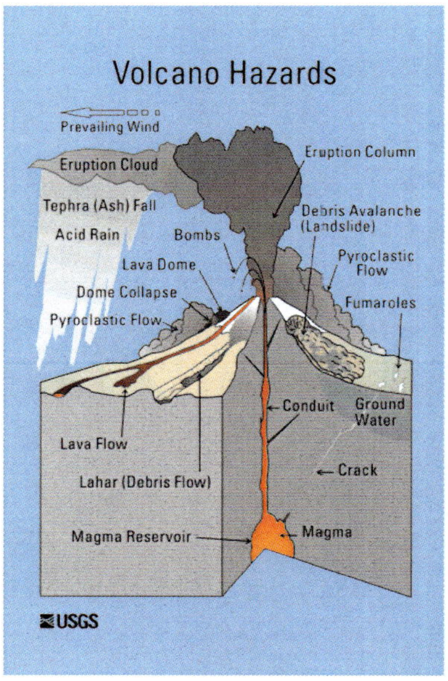

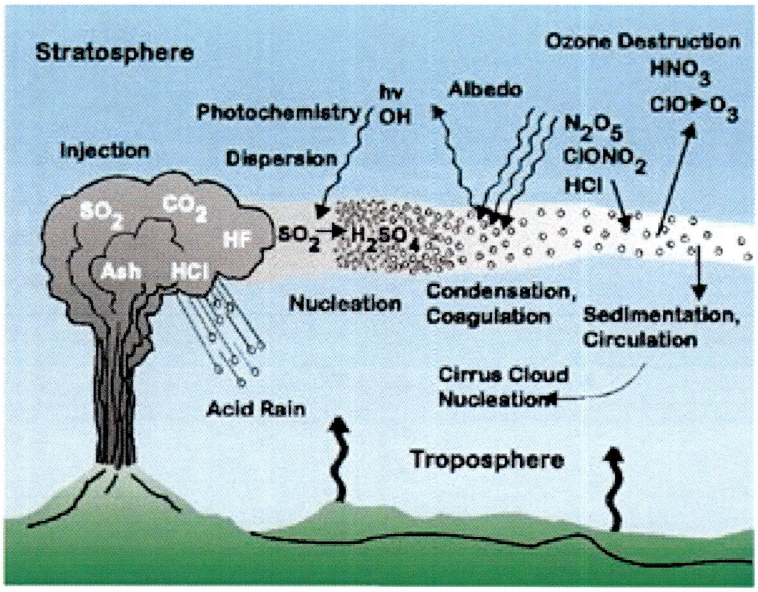

Many different gases make up ***volcanic gases***. The main ones are water vapor (H_2O), carbon dioxide (CO_2) sulfur dioxide (SO_2). ***hydrogen sulfide*** (H_2S), nitrogen, argon, helium, neon, methane, carbon monoxide (CO), and hydrogen. In addition, you can find meteoric oxygen, hydrogen chloride (HCl), hydrogen fluoride (HF), hydrogen bromide (HBr), nitrogen oxide (NO), sulfer hexafluoride, carbonyl sulfide, and organic compounds. Trace chemicals include mercury, halocarbons (including CFC's), and halogen oxide radicals. In other words, a lot of dangerous stuff is ejected from volcanic eruptions!

Not only are there immediate hazards from volcanic eruptions, but terrible scars are left on the earth in the aftermath. The following picture demonstrates some of these scars of volcanic eruptions.

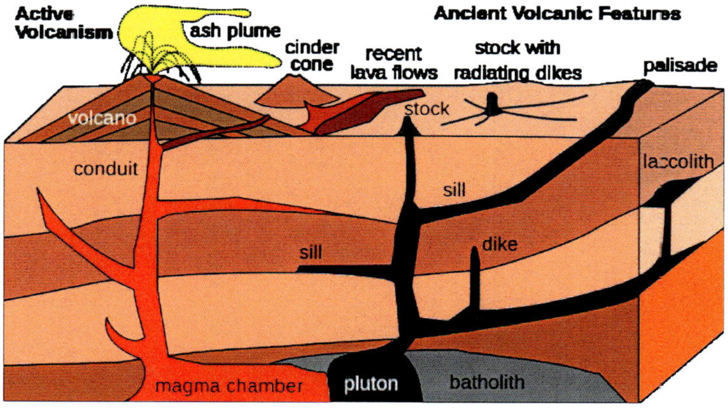

In summary, volcanic hazards destroy and incapacitate the landscape. They do not create an environment suitable for evolution to take place.

Of course, God watches over His earth and continues to care for the people and the planet. Volcanic rock breaks down to form and nourish soil. And in a sense the earth is partially renewed. This can be seen in the aftermath of the Mt. St. Helens eruption. What was once a barren landscape has regained some of the beauty that reflects the Creator.

Below is a list of the potential destructiveness of volcanic eruptions.

1. CO_2 emission rates from volcanoes have been estimated at between 143 million and 485 million tons annually.[16] However, the evidence reveals that volcanic eruptions were much greater in the past. How would this square with evolving life?
2. Oxygen would have been part of the equation and necessary for life, but oxygen also destroys developing DNA.
3. Volcanic eruptions emit hot ash which is initially deadly. The ash would have been extremely dangerous to air breathing animals. Mt. St. Helens alone, a small volcano, produced 540 million tons of ash in nine hours over 22,000 square miles.[17] Multiply this by 50,000 (estimated number of volcanoes in the past) and that would spell disaster to developing life. Take note of the recent eruptions in Iceland, where much of Europe was shut down in the aftermath of the eruptions. And these eruptions did not even begin to approach those of the past.
4. Volcanic eruptions emit sulfur dioxide. It has been charted between 1.6 million tons and 55 million tons per year (depending on the volcanic eruptions of the particular year). Again, multiply this by 50,000 and that would be deadly to developing life. Not only this, but $S0_2$ combines with water to form sulfuric acid.[18]
5. Evidence would indicate that a global ice event ("ice age") was brought on by the massive volcanism that likely occurred during and after the Flood.
6. Volcanoes emit hydrogen sulfide as well as the acidic gases hydrogen chloride and ***sulfur dioxide.*** These latter two irritate the eyes and throat. All three would also be harmful to developing life from an evolutionary perspective. In addition, lakes and other bodies of water dissolve these gases and the water becomes acidic. This water would be so acidic that it could burn a swimmer's skin in minutes.
7. Volcanoes emit hydrogen fluoride, a poisonous gas strong enough to etch glass!
8. In the evolutionary scheme of things, the formula for the first evolving life would be *time + self-organizing matter + volcanic eruptions + self-organizing biological molecules = life.* The main creative force was the volcanoes or fire. Isn't it interesting that in 2 Peter 3:5 the creative force is God, the earth being formed out of water and by water, and the destructive force in the end will be fire.

[16] *Volcanoes Can Affect Earth's Climate,* article found at https://volcanoes.usgs.gov/vhp/gas_climate.html.

[17] *How Much Ash Was There From the May 18, 1980 Eruption of Mt. St. Helens?* Article found at https://www.usgs.gov/faqs/how-much-ash-was-there-may-18-1980-eruption-mount-st-helens?qt-news_science_products=0#qt-news_science_products.

[18] *Volcanoes Can Affect Earth's Climate,* article found at https://volcanoes.usgs.gov/vhp/gas_climate.html.

Activity #7
Learning about Volcanic Hazards and Emissions

Read two articles about one or two recent volcanic events and write a paper of 500 words or more on what you learned about volcanic hazards.

Take Quiz 7

Lesson Eight – Geysers and Other Thermal Features

Words to know: geyser fumarole hot spring thermal thermophile geyserite travertine silica limestone pressure mud pots black smoker white smoker precipitate sulfide supervolcano sinter sulfuric acid

Just about everyone has heard of Yellowstone National Park. It is America's first national park, being signed into law in 1872 by President Ulysses S. Grant. Today Yellowstone National Park occupies 3,468 square miles in mostly in northwestern Wyoming and sections of Montana and Idaho. The Park is known for its animals, ***thermal*** features and its waterfalls.

The thermal features of Yellowstone National Park are the indirect result of the past volcanic activity of the Yellowstone Caldera. The combination of magma underneath the earth's surface, earthquakes, the rhyolite lava rock built up from the caldera's past eruptions, and the topmost layer of gravel deposited on top of the rhyolite lava by the Ice Age all contribute to the present system of thermal features of Yellowstone Park.

It has only been recently, however, that Yellowstone Park has been considered the result of a huge volcanic eruption from a type of volcano called a caldera. A caldera is so large, that its outlines can only be seen from the air. Below is what geologists think happened when Mt. Mazama, also a caldera, exploded.

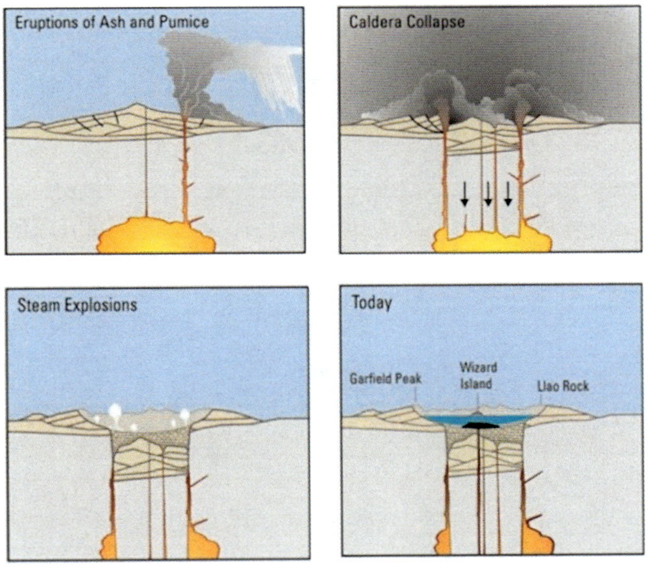

Stages in a Caldera Eruption
1. Magma builds under pressure and explodes in ash and pumice. 2. The caldera collapses under the weight of the lava and cracks in the crust. 3. This is followed by steam vents which are preserved as thermal features today.

How old is the Yellowstone Caldera (***supervolcano***)? That is an important question and one we will have to address if we are going to understand the geology of the Genesis Flood. Geologists tell us that the last major eruption of the Yellowstone Caldera was over 600,000 years ago. Now it should become clear that according to Genesis the global flood of Noah's day occurred around 4,500 years ago. If the Yellowstone caldera erupted before the Flood, and the Flood was truly a global, historical event, then no evidence should be left of the explosion, having been wiped out by the Flood. Consequently, the Yellowstone eruptions can not be as old as secularists would say.

What scientific method do geologists use to come to the conclusion that the Yellowstone Caldera is that old? There are no dates on the preserved lavas. There are no rates of thermal activity that can be measured for 600,000 years. The only method that geologists use is radiometric dating. And we have already seen that it is not reliable. So, let's go back to our Scriptural historical account of Noah and try to come up with an approximate age for the Yellowstone Caldera.

1. The history of the earth begins with the very first verse in Genesis 1:1 – "In the beginning God created the heavens and the earth." According to the genealogy/chronology in Genesis chapter 5, that event took place about 6,000 years ago.
2. Again, according to the genealogy/chronology in Genesis, the Genesis Flood took place in the 600th year of Noah's life – about 4,500 years ago.
3. According to Genesis chapter 1:31 the earth was in perfect harmony with God's character and His work. We would not expect natural disasters and death to have been a part of God's original creation.
4. Therefore, disasters like earthquakes and volcanic eruptions would not have been a part of the original creation, but of the Flood. The Yellowstone Caldera eruption would have taken place sometime during or shortly after the Flood year, about 4,500 years ago.

What kind of volcanic material did the Yellowstone eruption produce?
The Yellowstone eruption was an explosive eruption producing the following:

a) Basaltic, andesitic and rhyolitic lava flows
b) Rhyolitic pumice, ash, and tuff flows

Basalt flow (columns)

Rhyolite flow

Rhyolite obsidian flow

Volcanic tuff

Volcanic ash from Yellowstone Caldera has been found as far as eastern Louisiana.

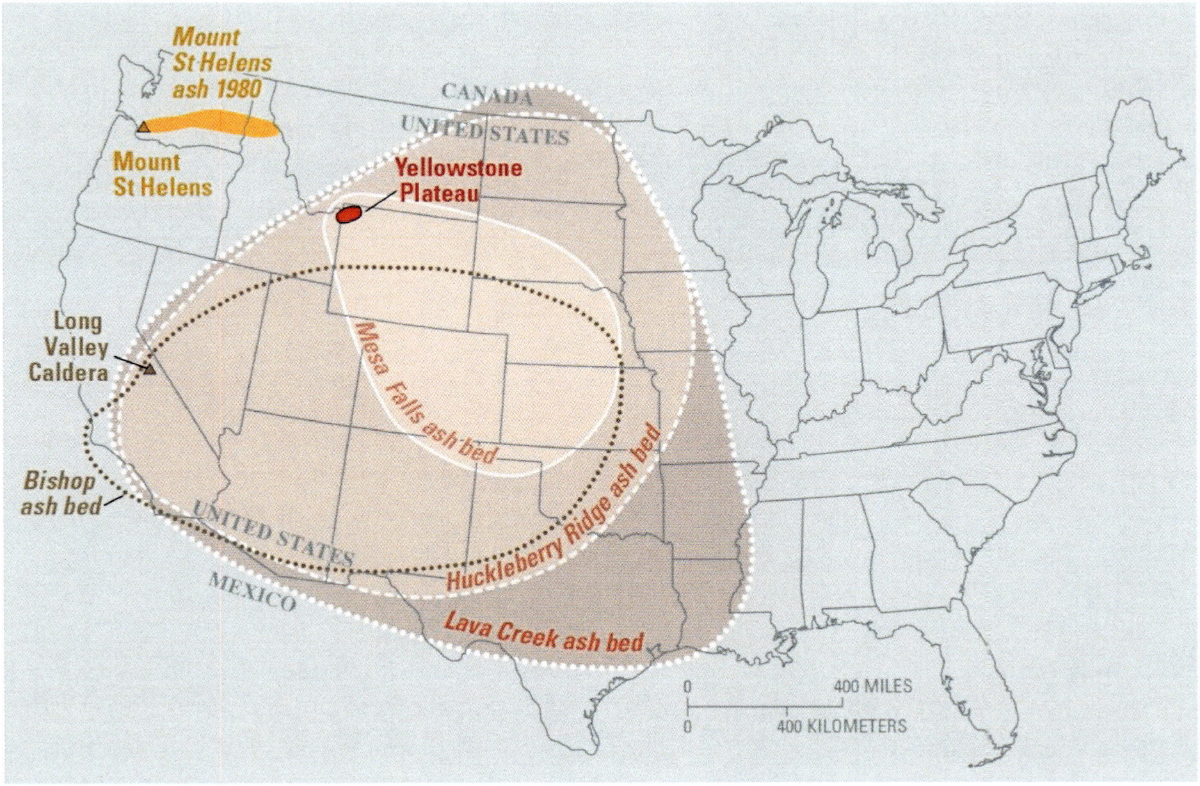

What thermal features now occupy the site of the Yellowstone Caldera?
There are four basic types of thermal features in Yellowstone Park all representing the ***pressure*** and magma underneath the ground at Yellowstone.

1. Geysers **–** the word geyser is an Icelandic word meaning *to gush.* Geysers range in size from little spouters all the way to a few hundred feet (eruptive height). There are two classes of geysers (a) those that erupt from pools and, (b) those that erupt from cones.

Fountain Geyser (eruption from a pool) and Beehive Geyser (eruption from a cone)

How does a geyser work?

In order for a geyser to work, there must be an adequate source of water and there must be a "plumbing system" through which the water travels with sufficient enough pressure and constriction to produce an eruption. In fact, this last factor is the main difference between a ***hot spring*** and a geyser – there isn't enough ***pressure*** or constriction and therefore the heated water rises but does not erupt.

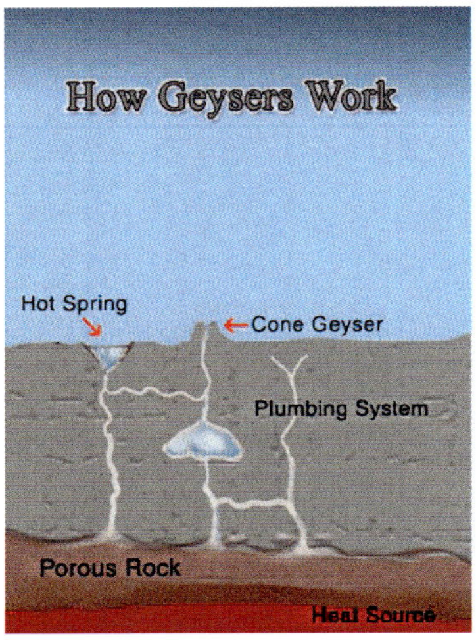

Water from below ground passes through the porous rhyolitic rock beneath Yellowstone. The water contains carbon dioxide which when mixed with the water produces carbonic acid. This in turn dissolves the ***silica*** in the rhyolite lava and carries it to the surface where it is deposited as a whitish-grayish rock called ***sinter*** or ***geyserite***. This end product is known as a ***precipitate.*** It is known as a precipitate because it comes out of a solution and is deposited.

Sinter or geyserite

The structures at Mammoth Hot Springs in northern Yellowstone are formed in exactly the same way except that the hot water passes through ***limestone,*** dissolving it, and then redepositing it on the surface, creating a ***travertine*** structure. Travertine, then, is a limestone precipitate.

Fumaroles – is from a Latin word meaning, *smoke*. A fumarole is an opening in the earth's crust, often in the neighborhood of volcanoes, which emits steam and gases such as carbon dioxide, sulfur dioxide, hydrogen chloride, and hydrogen ***sulfide***. The steam is created when superheated water turns to steam as its pressure drops when it emerges from the ground. Often this action is accompanied by a hissing sound.

Hot Spring - a spring that is produced by the emergence of geothermally heated groundwater from the earth's crust. There are geothermal hot springs in many locations all over the crust of the earth. They are particularly abundant in Yellowstone Park.

Mud Pots - a sort of acidic hot spring, or fumarole, with a limited water supply It usually takes the form of a pool of bubbling mud. The acid and microorganisms decompose surrounding rock into clay and mud. An acid spring produces hydrogen sulfide gas, the peculiar smell of rotten eggs.

Microorganisms convert the hydrogen sulfide to energy and the result is ***sulfuric acid*** which dissolves the geyserite surrounding the spring, turning it to mud. These mud pots can be more watery at different times of the year depending on the water supply. When the water source is low, mud develops.

The color in the thermal features

You may be wondering about the beautiful colors often a part of the thermal features of Yellowstone. They are the result of two factors: (a) microorganisms called ***thermophiles*** and (b) mineral deposits. Thermophiles are living animals that thrive in high temperatures. The different colors can represent different thermophiles.

Thermophiles indicate a general temperature of the hot water in Yellowstone.

The color of the travertine and geyserite often indicate the presence of iron compounds.

Will the Yellowstone supervolcano erupt again? No one knows for sure. The official park literature states that it is unlikely for it to erupt, for at least the next 10,000 years.

Two observations would help clarify the answer to this question.

- The source for the eruption no longer exists – The Genesis Flood; the breaking up of the fountains of the great deep.
- Volcanic eruptions have slowly been subsiding in intensity throughout the years

The best guess is that although Yellowstone is considered to be an active volcano, what we see today are the remnants of a very destructive time in Earth's history.

Black and White Smokers – geysers, hot springs, fumaroles and mud pots are known as hydrothermal vents. Similarly, on the ocean bottom near cracks in the ocean floor are similar hydrothermal vents known as *black* or *white smokers*. Extremely hot water is released through these vents in near freezing ocean waters. This hot water is filled with minerals that precipitate out to form cones, similar to geyser formation. This water is rich in dissolved minerals from the crust, most notably ***sulfides.*** When it comes in contact with cold ocean water, many minerals precipitate, forming a black, chimney-like structure around each vent. The deposited metal sulfides can become massive sulfide ore deposits in time.

White smoker vents emit lighter-hued minerals, such as those containing barium, calcium and silicon. These vents also tend to have lower temperature plumes. ***Black smokers*** were originally discovered in 1977. Black smokers are now known to exist in the Atlantic and Pacific Oceans, at an average depth of 6,300 feet. The most northerly black smokers are a cluster of five named Loki's Castle, discovered in 2008. It is not unusual to find teeming ocean life in colonies near these vents. This is most likely due to the favorable temperatures and minerals for certain forms of life.

Black and white smokers

Activity #8

The amount of volcanic material produced by Yellowstone Caldera compared with other volcanoes

Instructions: Choose six major volcanoes that have been active at one time or another in Earth's history. Construct a chart with columns with the following headings starting with Yellowstone Caldera:

- Type of volcano
- Type of lava produced
- Amount of pyroclastic material produced
- Types of hazards either produced by or estimated to have been produced by the volcanoes

Take Quiz 8

Lesson Nine – Earthquakes and Volcanoes

Words to know: tremor seismic fault tectonic Richter scale tsunami scarp escarpment epicenter lithosphere mantle core

Do you ever wonder why we have earthquakes today? If God created a very good earth, then why are there so many earthquakes on our planet? Earthquakes are destructive and result in misery, loss of life and loss of property.

Earthquakes remind us of the results of man's sin. It was because man had rebelled against God that the fountains of the great deep were opened and every land-dwelling, air-breathing animal and man outside the ark was killed. Not that each earthquake is God's punishment for a certain person's sin, but that the earth has been ruined because man was so evil.

What is an earthquake?

Take a look at the map of the world below. On it are recorded the earthquakes that have taken place from 1963-1998. Notice that the earthquakes are all centered around the cracks in the earth brought about by the breaking up of the fountains of the great deep.

Preliminary Determination of Epicenters
358,214 Events, 1963 - 1998

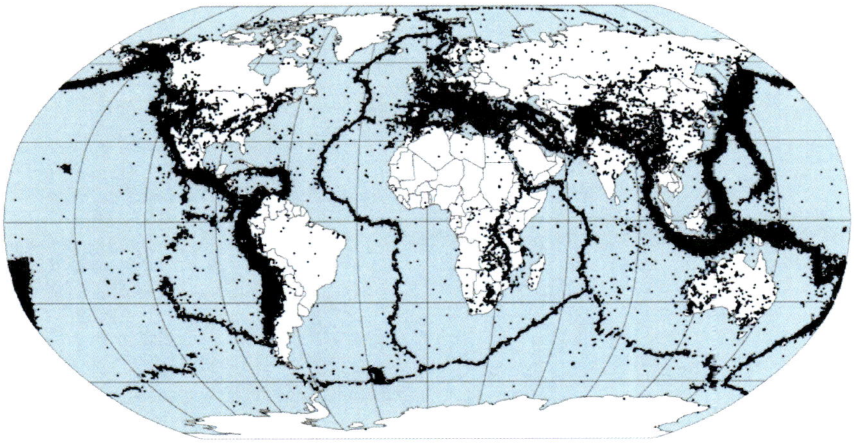

An earthquake is brought about by instability in the earth's crust. Because the earth was broken open during the 600th year of Noah's life, it is fragile and is constantly shifting. Geologists call this earth movement ***tectonic*** activity. Tectonic is an interesting word. The word is a Greek word meaning construction or carpenter. Secular geologists use the term to mean that the earth is constantly in a state of building or constructing itself up as in mountain building. This should remind us of another construct that secular geologists use – The Rock Cycle. They view Earth history as a cyclical process of building and recycling Earth materials to build our planet. Secular geologists use this term in a constructive way. Creationists should see earthquakes as destructive geologic events, not building events! Because secular geology has rejected the historical account of the Genesis Flood, they must be able to explain geologic processes as due to some other cause. They are constantly looking for an explanation for geologic processes that does not include the Bible. It is a big mistake and one that has led to over 250 years of blindness regarding Earth history.

An earthquake is also known as a quake, temblor, or ***tremor***. The seismicity, seismism or ***seismic*** activity of an area refers to the frequency, type and size of earthquakes experienced over a period of time. *Seismometers* are instruments that measure motions of the ground, including those of seismic waves generated by earthquakes, volcanic eruptions, and other seismic sources. *Seismograph* is another Greek word meaning to draw seismic activities. It is often used to mean seismometer, though it is more applicable to the older instruments in which the measuring and recording of ground motion were combined, than to modern systems, in which these functions are separated. Both types provide a continuous record of ground motion.

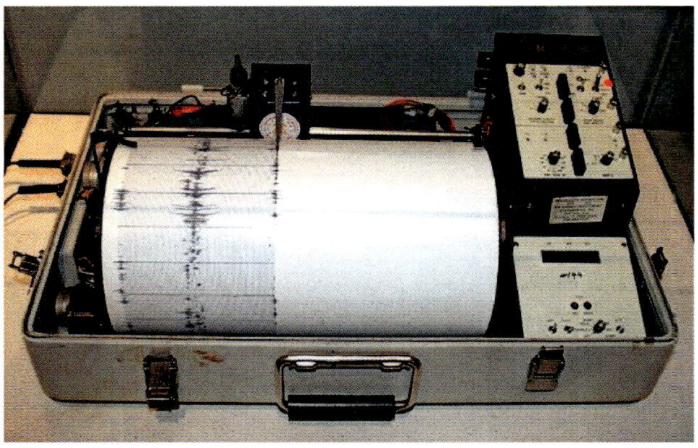

Breaks in the continuity of the earth's crust are called faults. Below is a picture of the three main types of faults. Faults consist of at least two sections of crust that can move in any direction. A reverse fault occurs when two blocks of rock are forced together by compression, and one section (or "hanging wall") moves upward. Contrary to that, a normal fault occurs when two blocks of rock are pulled apart, and one section ("hanging wall") moves downward. A strike-slip fault occurs when the fracture is predominantly horizontal, moving parallel to the fault.

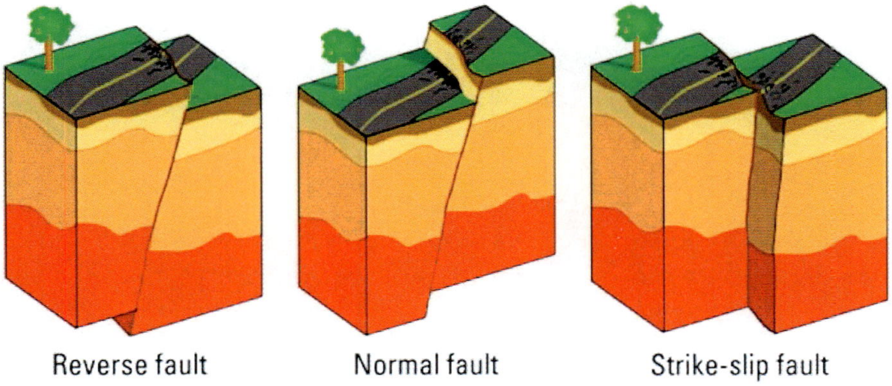

An ***esscarpment*** or ***scarp*** is the feature created by a fault. It expresses the distance between two sections of the earth where they have moved apart, generally because of an earthquake.

1959 Hebgen Lake Earthquake, Montana – escarpment, the distance between the two pieces of earth that have moved relative to one another.

A rift is where the earth's crust and ***lithosphere*** are being pulled apart, creating linear cracks. These are generally huge areas with lots of evidence of volcanism. An example of a rift is the Great African Rift. It is essentially a trench, approximately 3,700 miles in length, that runs from northern Syria to central Mozambique in South East Africa.[19] This is huge! What could have caused such a huge split in the earth? Genesis 7:11 explains it very well. Another example of a rift is the Mid-Atlantic Rift in Iceland. See ***Lesson 2.***

[19] Merriam-Webster, Inc., (1997). Merriam Webster's Geographical Dictionary 3/E(H). Merriam-Webster. p. 444.

Jordan Rift Valley, a part of the Great African Rift

The Interior of the Earth

How are rifts, faults, and volcanoes related to the interior of the earth? Does anyone know for sure what the interior of the earth looks like? Secular geologists guess as to what the interior of the earth looks like, but we don't know for sure. Below is an illustration of what most geologists think:

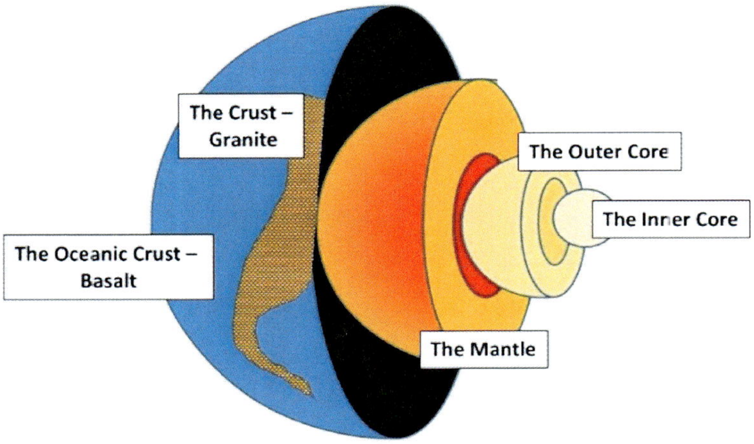

The upper portion of the earth's *crust* is the most easily accessible. But below that it is not known for sure what exists. Geologists think that there is a ***mantle*** below the crust. Geologists have added a new part to the mantle, called the asthenosphere, from the Greek, meaning weak sphere. This part of the earth is thought to be "plastic," made of partial magma melt. Below that geologists think there is a solid outer and inner ***core*** made of iron and nickel. But again, we don't know for sure. For now concentrate on knowing the four parts labeled above. Also know that the crust is apparently the thinnest part of the earth. It is called the lithosphere from the Greek, meaning *rocky sphere or shell.* It is the part we are most interested in, as that is where we find most of the rocks and fossils. According to best guesses, the shell or crust of the earth is only about 35 miles thick at its thickest. If that is true, the crust then comprises only a fraction of the total volume of the earth.

Both volcanoes and earthquakes represent the weaknesses produced by this breaking up of the fountains of the great deep. As earthquakes are associated with breaks or weaknesses in the earth, these coincide with volcanic activity. Although we do not know exactly what is under the crust of the earth, we do know that temperatures get hotter the further down we go. Seismic readings indicate a possibility of moving magma.

By way of example, in the early 20th Century geologists attempted to drill down into the earth's crust at Norris Geyser Basin in Yellowstone National Park. This geyser basin is the hottest surface area in Yellowstone because of the volcanic activity. After drilling a little over 200 feet, they had to stop. The temperature was over 400 degrees and would no long permit drilling with the equipment they had. Obviously, pressure and heat increase with depth.

The Epicenter of an Earthquake

When an earthquake happens, it is thought that there is a focal point where the initial movement of the earthquake takes place, followed by waves of energy travelling in all directions. The epicenter is that point vertically above the focus, where the fault begins to rupture, and in most cases, it is the area of greatest damage.

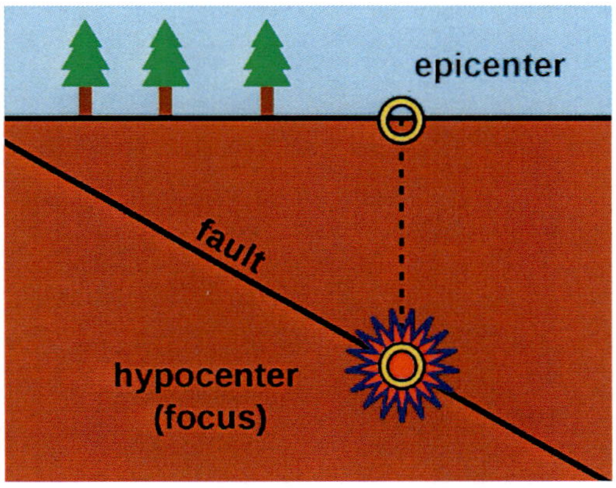

The Richter Scale

Developed in 1935 by Charles Francis Richter in partnership with Beno Gutenberg, both from the California Institute of Technology, the Richter scale was firstly intended to be used only in a particular study area in California, and on seismograms recorded on a particular instrument, the Wood-Anderson torsion seismograph. The Richter scale measures the energy released by an earthquake. Another scale, the Mercalli intensity scale, classifies earthquakes by their effects, from detectable by instruments but not noticeable, to catastrophic. The energy and effects are not necessarily strongly correlated; a shallow earthquake in a populated area with soil of certain types can be far more intense than a much more energetic deep earthquake in an isolated area. An earthquake with a 3.5 on the Richter scale is obviously not as intense as one measuring 6 or 7 (the most severe being 10).

Tsunami **–** is from the Japanese for *harbor wave*. A ***tsunami*** is a series of water waves caused by the displacement of a large volume of a body of water, generally an ocean or a large lake. There are many events that can trigger a tsunami, including earthquakes, volcanic eruptions, landslides, glacier calvings, and meteorite impacts.

Tsunamis are sometimes referred to as tidal waves, which are unusually high sea waves that are triggered especially by earthquakes. In recent years, this term has fallen out of favor, especially in the scientific community, because tsunamis actually have nothing to do with tides. The once-popular term derives from their most common appearance, which is that of an extraordinarily high tidal wave. Tsunamis and tides both produce waves of water that move inland, but in the case of tsunami the inland movement of water is much greater and lasts for a longer period, giving the impression of an incredibly high tide.

Tsunami damage

Some Destructive Earthquakes in the Past

Date	*Location*	*Deaths*	Magnitude	*Comments*
January 23, 1556	China, Shansi	830,000		
October 11, 1737	India, Calcutta**	300,000		
December 26, 2004	Sumatra	280,000	9.1	tsunami (animation; NOAA); about dozen countries affected
July 27, 1976	China, Tangshan	255,000	8.0	*
August 9, 1138	Syria, Aleppo	230,000		
May 22, 1927	China, near Xining	200,000	8.3	Large fractures
December 22, 856+	Iran,Damghan	200,000		
December 16, 1920	China,Gansu	200,000	8.6	Major fractures, landslides
March 23, 893+	Iran, Ardabil	150,000		
September 1, 1923	Japan, Kwanto	143,000	8.3	Great Tokyo fire
September, 1290	China, Chihli	100,000		
November, 1667	Caucasia, Shemakha	80,000		
November 18, 1727	Iran, Tabriz	77,000		
December 28, 1908	Italy, Messina	70,000	7.5	Deaths from 70,000 to 100,000 (estimated); earthquake and tsunami
November 1, 1755	Portugal, Lisbon	70,000	8.7	Great tsunami
December 25, 1932	China, Gansu	70,000	7.6	
May 31, 1970	Peru	66,000	7.8	great rock slide, floods
1268	Asia Minor, Silicia	60,000		
January 11, 1693	Italy, Sicily	60,000		
May 30, 1935	Pakistan, Quetta	30,000	7.5	Quetta almost completely destroyed
February 4, 1783	Italy, Calabria	50,000		
June 20, 1990	Iran	50,000	7.7	Landslides

Largest Magnitude Earthquakes of past 100 years

Location	*Date*	*Magnitude*[2]
Chile	May 22, 1960	9.5
Prince William Sound, Alaska	March 28, 1964	9.2
Andreanof Islands, Aleutian Islands	March 9, 1957	9.1
Kamchatka	Nov. 4, 1952	9.0
off the coast of Sumatra	Dec. 26, 2004	9.0

Off the coast of Ecuador	Jan. 31, 1906	8.8
Rat Islands, Aleutian Islands	Feb. 4, 1965	8.7
India-China border	Aug. 15, 1950	8.6
Kamchatka	Feb. 3, 1923	8.5
Banda Sea, Indonesia	Feb. 1, 1938	8.5
Kuril Islands	Oct. 13, 1963	8.5

Activity #9(a)
Learning about Earthquakes

(a) Instructions: Do some research on earthquake preparedness. Construct a chart of the various situations to prepare for, and how to prepare for them in the event of an earthquake. Put together a kit for earthquake preparedness.

Activity #9(b)
Learning about Earthquakes

(b) Instructions: Follow the internet link (or a similar link) below to the city of San Francisco and the earthquake of 1906. After you have read through the information and toured the museum site on line, fill out the chart below. Use additional paper if necessary.

Virtual Museum of the city of San Francisco - http://www.sfmuseum.org/1906/06.html

Earthquake Information Chart

When did the earthquake take place?	
Where did the earthquake happen? Describe the places, the people who live there, and the types of buildings that are found there. What happened to the places during the earthquake?	
Why did the earthquake happen? Describe the type of land formations in the area. Could the earthquake be related to these land formations?	
Describe the recovery efforts, and how people are preparing for possible future earthquakes.	

Take Quiz 9

Lesson Ten – Volcanoes, the Genesis Flood and the Ice Age

Words to know: *erratics* *striations* *horns* *cirques* *volcanic aerosol* *glacial trough/valley*

Why don't secular geologists today accept the Genesis Flood? Is it really because it is not science, as they claim? One of the lessons we can learn from the Enlightenment and the overthrow of Biblical geology in the 1800s is that the Genesis Flood was never *disproved scientifically*; only rejected!

The Enlightenment gave us the religion of Deism – *God probably created the heavens and the earth, but He has not and does not intervene in its affairs.* This is not a scientific statement, but a religious one. Thus, the Enlightenment was a worldview collision with the Biblical worldview.

Psalm 29:10 states, "The Lord sat as King at the flood; yes, the Lord sits as King forever." The reaction during the Enlightenment to anything we cannot see or measure was a *knee-jerk* reaction. It not only rejected superstitions and the alchemy of the time, but also rejected the supernatural. It rejected the Lord as King throughout all time and so drifted into darkness and embraced a false worldview of naturalism – only what could be tested by man would be accepted as science and therefore legitimate. What a shame.

We have seen the power, the destruction and the hazards of volcanoes. Could they really have been part of the creative work of God in Genesis chapter 1? After all, when God was finished creating, He pronounced everything that He had made as "very good." It makes more sense to place the origin of volcanoes during the time of the global flood of Genesis when "...the fountains of the great deep burst open...." Volcanoes therefore serve as reminders that God holds man accountable to His values. Volcanoes also serve as reminders of the destructive consequences of man's rebellion against God.

One of the most significant geologic results of the volcanic rifts caused by the Genesis Flood could have been the so-called Ice Age. The geologic evidence that much of the earth has been covered by a lot of snow and ice in the past is irrefutable. The big question is not if it happened, but when it happened. How does it fit into the Biblical chronology? And what part could volcanoes have played in the Ice Age?

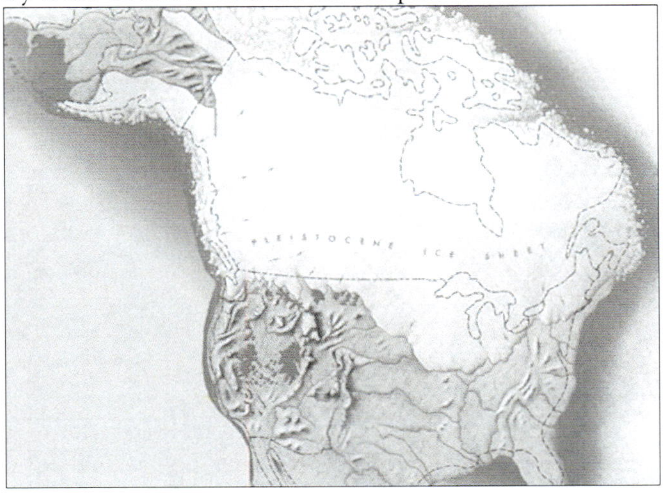

Secular geologists are able to show through the evidence that an ice age occurred in Earth's past. There is disagreement on the number of ice ages, the exact dates and the causes. But there has been a lot of ice and snow that covered the Northern Hemisphere in particular. Notice the map below to see the estimated ice coverage during this mysterious time in Earth's history.

How do geologists know this? Here is some of the evidence for an ice age:

- ***Glacial erratics*** – boulders that are out of place; they don't belong in the country in which they are found. The boulders in Northern Yellowstone Park are made of granite. The predominant rock in Yellowstone is volcanic.

- ***Glacial striations*** – scratch marks made on hard rock surfaces, the result of ice moving rock over the surface.

- ***Glacial horns and cirques*** – huge amounts of rock carved and sculpted where glaciers have formerly been. These indicate the height and amount of ice coverage. There is evidence that Yellowstone National Park was covered by as much as 3,500 feet of snow and ice. But there are no glaciers in Yellowstone today.

Glacial horns

The curved bowl is a glacial cirque.

- ***Glacial troughs or valleys*** – massive rounded troughs that have been carved by moving glaciers in the past. These glaciers are no longer present in most places, but evidence of their incredible size has been left behind. This evidence points to a single event in the not too distant past when there was a rapid build-up of ice, followed by rapid melting. Uniformitarian explanations cannot account for this glacial activity. Most likely it was due to the profuse volcanic eruptions during, and for awhile shortly after, the Flood.

This glacial trough is in the Beartooth Mountains, part of the Rocky Mountains, in Montana. The glacier that carved this is no longer present, but its evidence is pronounced.

What do volcanoes have to do with an ice age? Study the picture below. There are two main points you will want to get from this diagram:
1. The effect of ash cloud coverage on the sun's reflective energy
2. The net effect of cooling of the earth's land mass as a result

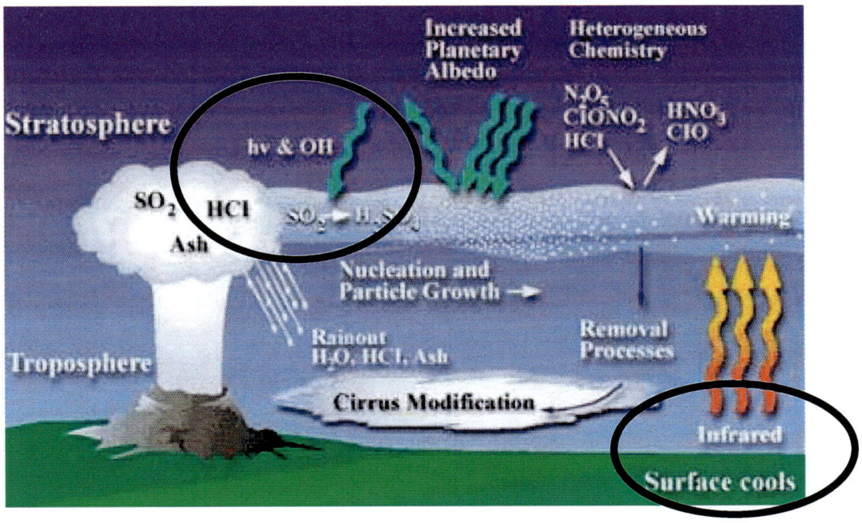

There is one other point not covered in the diagram. Where would the moisture come from that would have produced an abundance of snow and ice? What would cause an increase in evaporation such that the cooled atmosphere would turn it to lots of snow and ice? This has been one of the main challenges to a uniformitarian model for explaining an ice age.

We don't get an abundance of snow in cold weather. To produce the kind of moisture required for there to be a long time period of continuous snow and ice buildup would require a unique geologic event. Secular geology through the years has compiled a list of over 100 ideas about how an ice age would have started. Not one of those ideas has been totally accepted. And no one idea has been able to explain how an ice age would have come about. There is still much disagreement in the secular geology community on what could have triggered the onset of an ice age. But the Genesis Flood provides the mechanisms necessary for producing such a time in our history. It is the unique combination of these mechanisms that would have brought on an Ice Age.

- With the breaking up of the fountains of the great deep would have come a release of magma thus warming the ocean waters as they covered the earth
- With the breaking up of the fountains of the great deep would have come ruptures in the earth's crust bringing on lava flows and volcanic eruptions.
- With the tremendous amount of volcanic activity would have come an atmosphere filled with ***volcanic aerosols*** which would have reduced the sun's reflective energy, thus cooling the atmosphere.
- Warmer oceans would have produced more evaporation.
- Increased moisture in a cooler atmosphere would have produced a lot of snow very quickly.
- With increased snow depth and pressure would have come ice.

The combination of these factors, brought on by increased volcanism, would have resulted in a rapid buildup of snow and ice, particularly in the Northern Hemisphere and in the higher latitudes. An Ice Age following the Genesis Flood, lasting from between 500-700 years would have been the result.

Activity #10

Understanding the Ice Age from a Biblical Perspective

Instructions: construct an events line showing the various stages of an ice age build up and meltdown using the information from your lesson.

Take Quiz 10

Lesson Eleven: Mt. St. Helens and its Lessons for a Biblical Understanding

Words to know: *atomic bomb* *strata* *stratigraphy*

Lessons from the Mt. St. Helens Eruption

Mt. St. Helens has been the most studied volcano in history. Volcanology took huge leaps as a geologic discipline after the eruption of Mt. St. Helens in 1980. Mt. St. Helens is a study in the power and effects of catastrophism.

What are some of the lessons from Mt. St. Helens for a Biblical understanding of the Genesis Flood?

1. Secular geologists are now more accepting of catastrophic geologic processes as a result. Previously secular geology was almost exclusively limited to slow processes acting over millions of years. Anything catastrophic was considered to be anathema in secular geology. It sounded too much like Noah's Flood.
2. Geologists have seen thick beds of ash deposited in less than an hour and display fine laminations. Previously these kinds of laminations would have been interpreted to mean that the layers took millions of years to form and that they represented past environments recorded in the layers.
3. Geologists saw deep canyons in the Mt. St. Helens area form quickly as the result of the catastrophic lahars and mud/ash flows from Mt. St. Helens. In fact, one of the canyons at Mt. St. Helens has been labeled, *The Little Grand Canyon.* In other words, it did not take millions of years for a river to carve a canyon. It could have happened quite rapidly under the right conditions.
4. Geologists saw rocky surfaces with grooves and striations that were scraped by rock blasted along the ground instead of the usual explanation that grooves and striations were caused by moving glaciers over a long period of time.
5. Geologists have also seen that the once-hallowed concept of radiometric dating was demonstrated to be inaccurate and unreliable as the result of conflicting and old dates acquired from Mt. St. Helens lava geologists knew to be less than 10 years old.

Below is an aerial map of the Mt. St. Helens area. The blown-out portion faces due north.

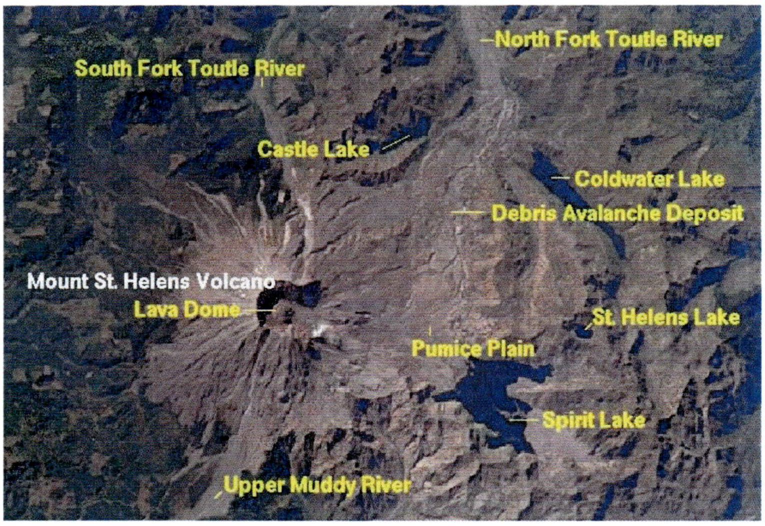

N

How old is Mt. St. Helens?
As we have seen in a previous lesson, Mt. St. Helens is part of The Ring of Fire. Therefore it must have been formed as part of the breaking up of the fountains of the great deep in Genesis 7:11. So, Mt. St. Helens is about 4,500 years old. Geologists say it is about 40,000 years old. How did they get that? Remember that there are no dates or ages imprinted on the rocks themselves. The age must be interpreted from some other source. For the Christian that interpretation comes from their Scriptures. For the secular world the interpretation comes from ***(a)*** the belief that the earth must be older than what the Bible says, and, ***(b)*** backed up by radiometric dating. Remember too that radiometric dating is itself an interpretation of radioactive decay.

The main thing we have learned from the eruption of Mt. St. Helens is that catastrophic processes do not take a long time to sculpt and shape the earth. Here are just a few of these processes:

1. The volcano destroyed a beautiful mountain in a matter of minutes. In addition, the abrupt release of pressure over the magma chamber created "a glowing cloud of superheated gas and rock debris blown out of the mountain face moving at nearly supersonic speeds. Everything within eight miles of the blast was wiped out almost instantly, according to USGS. The shockwave rolled over the forest for another 19 miles, leveling century-old trees; all the trunks neatly aligned to the north. The area devastated by the direct blast force covered an area of nearly 230 square miles (596 square kilometers) .... The heat of the initial eruption melted and eroded the glacial ice and snow around the remaining part of the volcano. The water mixed with dirt and debris to create lahars, or volcanic mudflows. According to USGS, the lahars reached speeds of 90 mph (145 km/h), and demolished everything in their path. Most of the glaciers surrounding Mount St. Helens melted...."[20] The Genesis Flood was almost certainly accompanied by many volcanic eruptions such as Mt. St. Helens. How much rearrangement of the earth do you think was accomplished in the first few months of the Genesis Flood? Look at the picture on the next page and notice just how much destruction was accomplished by just one eruption. What could be mistaken for millions of years of erosive processes was accomplished in just a few short minutes!

Mt. St. Helens – before and after

[20] *Mt. St. Helens Eruption: Facts and Information*, Mary Bagley, ***Live Science*** October 16, 2018, article found at https://www.livescience.com/27553-mount-st-helens-eruption.html.

2. Canyons were formed in just a few short months. One such canyon at Mt. St. Helens has been called, *The Little Grand Canyon*. The force of lahars and volcanic landslides would have worked quickly on the soil.

One of the many canyons formed at Mt. St. Helens

3. Features very similar to the Badlands of South Dakota formed in just a few days. The Badlands of South Dakota, too, are filled with eroded soils and ash, indicating catastrophic formation. The Badlands are considered to be between 23-25 million years old, yet they look very similar to the topography of Mt. St. Helens.

Mt. St. Helens aftermath

The Badlands of South Dakota

4. The various ash flows from Mt. St. Helens formed layered ***strata*** within just a few hours. Normally layered strata are interpreted by secular geologists as records of past environments that were preserved over millions of years. Measuring from a fraction of an inch to a yard in thickness, each of these strata took from a few seconds to a few minutes to form. ***Stratigraphy*** is the science of reading the strata.

5. The eruption of Mt. St. Helens produced a lot of catastrophic flooding due to snow melt mixed with mud, ash and volcanic rock. Drainage systems were rapidly formed within several hours.

Debris avalanche deposit

6. Log mats from the stripped forests around Mt. St. Helens produced a log jam. In just ten short years, over one million trees that washed into Spirit Lake had begun to show similar characteristics as the so-called "fossil forests" of Yellowstone Park. For many years, the official interpretation of these many fossil trees in Yellowstone National Park was that they were buried there over millions of years, layer by layer, under sediments At Mt. St. Helens, many of the trees knocked down by the eruption and buried by sediments in Spirit Lake were buried in an upright position. Had these been discovered without initially knowing that there had been a major volcanic eruption, these fallen and buried trees might have been interpreted as successive fossil forests. Today, the official position of Yellowstone National Park has been changed, due to what was discovered at Mt. St. Helens.

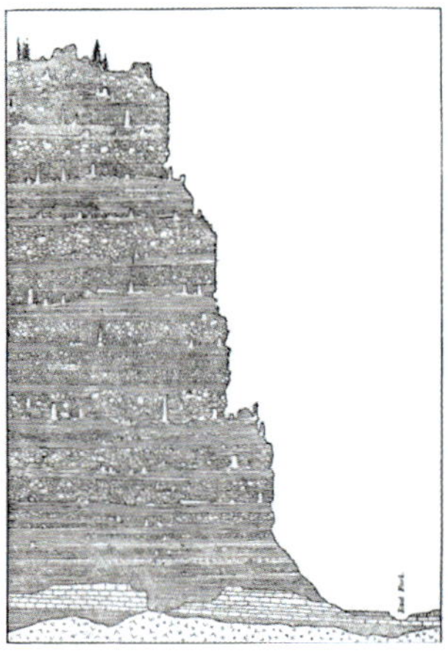

"Fossil Forests," Yellowstone, showing the original interpretation of thousands of years of successive forest growth, burial, and petrification; Fossil trees in upright position at Specimen Ridge; Log mat at Mt. St. Helens: people in boat circled for size

7. What is probably the most amazing event at Mt. St. Helens is just how quickly life came back to the area. What was once considered a totally and irreversibly damaged "moonscape" is now quickly renewing itself with an abundance of plant and animal life. God made the earth so that it could recover and renew itself. Both plants and animals quickly flourished and spread after the Flood. Today, we see just vestiges of once mighty volcanic eruptions, fossilization and quickly-formed strata. And this is the very same story that we see at Mt. St. Helens.

Final Project

Build a scale model of Mt. St. Helens before and after the 1980 eruption. Possible materials used could be: clay, chicken wire, papier mache, computer drawings.

Take Quiz 11

Appendix A

Plate Tectonics – This idea is increasingly gaining acceptance with secular and catastrophist geologists. So, we need to become somewhat acquainted with it.

What is Plate Tectonics? First of all we must realize that Plate Tectonics is an idea, not proven to be true by scientific method. And we also need to realize that Plate Tectonics is a uniformitarian idea. That means that it will rely on what we observe now to explain the past history of the earth. As such it denies that the Genesis Flood ever took place.

Plate tectonics is based on the idea that continents were all once connected as one large land mass but have been moving away from each other for millions of years. In the early 20th Century it was suggested that the continents looked as if they could fit together like the pieces of a jigsaw puzzle.

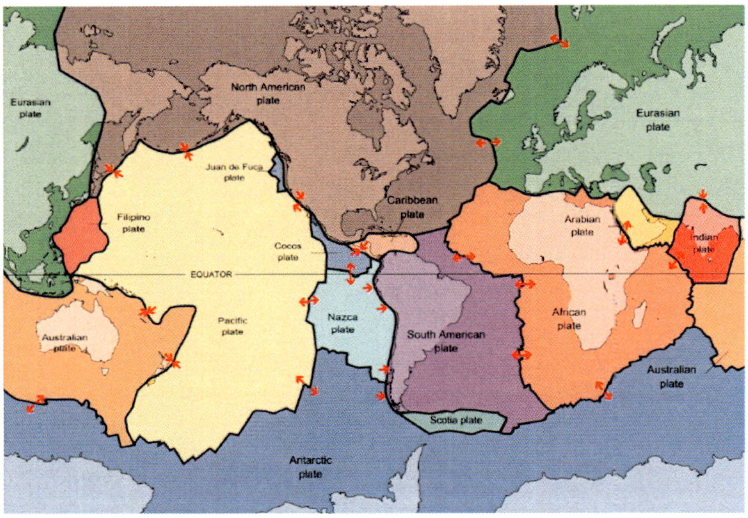

But at the time there were no mechanisms that could explain how such a feat could be accomplished. After all, the continents are not like floating pie plates on the ocean. They are anchored. Then in the 1960s and 1970s with advanced oceanography technology there were discovered huge cracks on the ocean floor. Some of these cracks were deep canyons, larger than the Grand Canyon. It was also confirmed that the ocean floor is made up of basalt lava with a veneer of sediment on top with basalt lava continuing to erupt. The idea of *sea floor spreading* developed which just simply states that it is a process that occurs at mid-ocean ridges, where new oceanic crust is formed through volcanic activity and then gradually moves away from the ridge. Now, here is where uniformitarianism takes over. Because the past is always interpreted by what can be seen to be happening now, the action of sea floor spreading was projected into the past at present rates to arrive at a mechanism for continental drift. If the Flood had never taken place, as modern geologists insist, then this explanation or any explanation for that matter would suffice to explain the movement in the earth's crust. But because a Christian trusts in the revelation of God, the Scriptures, he/she believes a different story. The Genesis Flood explains the rifts or ridges, the cracks, earth movement, volcanism, eruption of lava and the current trend in forming additional oceanic floor. It's all quite simple really. But you can see now how big of an impact the philosophy of uniformitarianism plays in modern thinking!

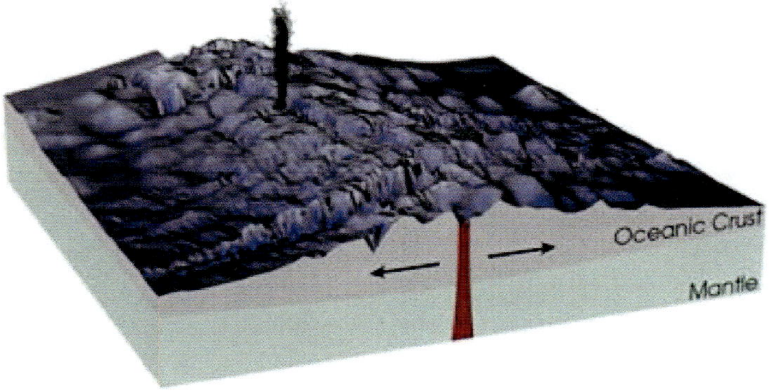

The idea of sea floor spreading is illustrated to the left. Active on-going volcanism would lead uniformitarian geologists o conclude that the sea floor has been spreading for millions of years.

Now, it is true that the earth exhibits cracks all around the globe on the bottom of the ocean floor. And we know that there are earthquakes that do move sections of rock and land relative to one another. These are called faults. But to say from this evidence that the continents have been moving for millions of years is stretching the idea way beyond science.

The Mid-Atlantic Rift runs right through the country of Iceland. This area spread slightly, but does not continue to slide apart. Iceland is a volcanically active

The above satellite and topographical maps show the Caucasus Mountain Range that divides the Black Sea from the Caspian Sea. Why is it there and how did it form? Since no geologist was there when this event happened or started to happen, there are really only two explanations. Both explanations require a prior commitment to a worldview:

1. Modern geology, with its rejection of a Biblical Flood and supernatural creation, would say that the Caucasus Mountains formed largely as the result of a tectonic plate collision between the Arabian plate moving northward with respect to the Eurasian plate. Further they would say that the formation of the modern Caucasus took millions of years. While the sea, the land, and strong volcanic processes were changing, the climate was changing too. Species of animals and plants continued to come into the world and then disappear. Geologists would place the age of these mountains around 30-40 million years ago, the area having been previously covered by water for 20 million years. Earth movement has indeed occurred being manifested in earthquakes and volcanic eruptions. But just how far the continents have moved or are moving apart or colliding is a matter of speculation based on a uniformitarian assumption.

2. Biblical creationists would say that during the year of the Genesis Flood the fountains of the great deep burst open causing cracks/splits in the earth out of which came hot water and magma. Thus, the evidence for volcanic and plutonic origin. Toward the end of the Flood, as Psalm 104:5-9 indicates, the mountains rose, and the valleys sank down causing rapid mountain building and sheet and channelized erosion. Thus, the Caucasus Mountain Range was born roughly 4,500 years ago. The rift as seen in the maps above is the remnant of this global flood event. The interpretation relies on a prior commitment to a worldview. The answer cannot be found in the evidence alone.

Appendix B -Dating Systems and the Age of the Earth

In modern geology, the only dating systems that are accepted within scientific circles are those which yield an old age of the earth in line with the now accepted 4.6 billion years. Any dating method that yield ages too old are frowned on. Any dating method that yield ages too young are flatly rejected! Anything with a young age for the earth would remind people of the Book of Genesis. Therefore, these ages are resolutely rejected from even being considered.

In the 1970's, the late Dr. Henry Morris, one of the pioneers of the Creation Movement, published a little book, entitled, ***The Scientific Case for Creation***. It is no longer in print. One of the tables he published in that little book is a compilation of various dating systems that yield ages for the earth that are all over the geological map! These systems have all been rejected because they do not line up with the accepted age of the earth. All these dating systems use present, observed processes which can be measured and recorded. Then uniformitarian assumptions are applied, as in all dating systems, to the presently observed process rates – just like those used in radiometric dating methods. So, comparing apples with apples, all these dating systems demonstrate that no system can give us conclusive scientific evidence that the earth is 4.6 billion years old! The following is the table that was published in the 1970s and it is still applicable for today's discussions. Note that the dating of the earth has been all over the map. Which dating system will you use? What is the standard you are going to use to determine which one is correct?

Table 1
Age of the Earth based on Uniformitarian Assumptions

(Unless otherwise noted, based on standard assumptions of closed systems, constant rates, and no initial daughter elements.)

	Dating System Used	Indicated Age of the Earth
1.	Efflux of Helium-4 into the atmosphere	1,750 – 175,000 years
2.	Influx of meteoritic dust from space	Too small to calculate
3.	Influx of radiocarbon to the earth system	5,000 – 10,000 years
4.	Development of total human population	Less than 4,000 years
5.	Influx of uranium to the ocean via rivers	10,000 – 100,000 years
6.	Influx of sodium to the ocean via rivers	260,000,000 years
7.	Influx of nickel to the ocean via rivers	9,000 years
8.	Influx of magnesium to the ocean via rivers	45,000,000 years
9.	Influx of silicon to the ocean via rivers	8,000 years
10.	Influx of potassium to the ocean via rivers	11,000,000 years
11.	Influx of copper to the ocean via rivers	50,000 years
12.	Influx of gold to the ocean via rivers	560,000 years
13.	Influx of silver to the ocean via rivers	2,100,000 years
14.	Influx of mercury to the ocean via rivers	42,000 years
15.	Influx of lead to the ocean via rivers	2,000 years
16.	Influx of tin to the ocean via rivers	100,000 years
17.	Influx of aluminum to the ocean via rivers	100 years
18.	Influx of carbonate to the ocean via rivers	100,000 years
19.	Influx of sulphate to the ocean via rivers	10,000,000 years
20.	Influx of chlorine to the ocean via rivers	164,000,000 years
21.	Influx of calcium to the ocean via rivers	1,000,000 years
22.	Leaching of sodium from continents	32,000,000 years
23.	Leaching of chlorine from the continents	1,000,000 years
24.	Leaching of calcium from the continents	12,000,000 years

	Dating System Used	Indicated Age of the Earth
25.	Influx of sediment to the ocean via rivers	30,000,000
26.	Erosion of sediment from continents	14,000,000 years
27.	Decay of earth's magnetic field	10,000 years
28.	Efflux of oil from traps by fluid pressure	10,000-100,000 years
29.	Formation of radiogenic lead by neutron capture	Too small to measure
30.	Formation of radiogenic strontium by neutron capture	Too small to measure
31.	Decay of natural remanent paleomagnetism	100,000 years
32.	Decay of C-14 in pre-Cambrian wood	4,000 years
33.	Decay of uranium with initial lead	Too small to measure
34.	Decay of potassium with entrapped argon	Too small to measure
35.	Influx of juvenile water to oceans	340,000,000 years
36.	Influx of magma from mantle to form crust	500,000,000 years
37.	Growth of active coral reefs	10,000 years
38.	Growth of oldest living part of biosphere	5,000 years
39.	Origin of human civilizations	5,000 years
40.	Formation of river deltas	5,000 years
41.	Submarine oil seepage into the oceans	50,000,000 years
42.	Decay of natural plutonium	80,000,000 years
43.	Decay of lines of galaxies	10,000,000 years
44.	Expanding interstellar gas	60,000,000 years
45.	Formation of Carbon-14 on meteorites	100,000 years
46.	Decay of short-period comets	10,000 years
47.	Decay of long-period comets	1,000,000 years
48.	Influx of small particles to the sun	83,000 years

	Dating System Used	Indicated Age of the Earth
49.	Maximum life of meteor showers	5,000,000 years
50.	Accumulation of dust on the moon	200,000 years
51.	Deceleration of earth by tidal friction	500,000,000 years
52.	Cooling of earth by heat efflux	24,000,000 years
53.	Accumulation of calcareous ooze on sea floor	5,000,000
54.	Influx of lithium into ocean via rivers	20,000,000 years
55.	Influx of titanium into ocean via rivers	160 years
56.	Influx of chromium into ocean via rivers	350 years
57.	Influx of manganese into ocean via rivers	1,400 years
58.	Influx of iron into ocean via rivers	140 years
59.	Influx of cobalt into ocean via rivers	18,000 years
60.	Influx of zinc into ocean via rivers	180,000 years
61.	Influx of rubidium into ocean via rivers	270,000 years
62.	Influx of strontium of ocean into rivers	19,000,000 years
63.	Influx of bismuth into ocean via rivers	45,000 years
64.	Influx of thorium into ocean via rivers	350 years
65.	Influx of antimony into ocean via rivers	350,000 years
66.	Influx of tungsten into ocean via rivers	1,000 years
67.	Influx of barium into ocean via rivers	84,000 years
68.	Influx of molybdenum into ocean via rivers	500,000 years
69.	Influx of bicarbonate into ocean via rivers	700,000 years
70.	Escape of high-velocity stars from globular clusters	40,000 years
71.	Rotation of spiral galaxies	200,000,000 years
72.	Accumulation of peat in peat bogs	8,000 years
73.	Accumulation of sediments for sedimentary rocks	20,000 years
74.	Lithification of sediments to form sedimentary rocks	20,000 years
75.	Instability of rings of Saturn	1,000,000 years
76.	Escape of methane from Titan	20,000,000 years

Quizzes

Quiz #1:

1. Who was the father of modern geology?
2. What does the word uniformitarianism mean?
3. What does the word deism mean?
4. What determines the viscosity of lava?
5. What does the word pyroclastic mean?
6. The Volcanic Explosivity Index refers to what?
7. An opening in the earth through which lava extrudes is called what?
8. Hot molten rock below the earth's surface is called what?
9. Briefly describe the difference between the secular view and the Biblical view of geology.

Quiz #2:

1. Name the six types of volcanoes.
2. What is the difference between the lavas of lava flows and the lava produced by stratovolcanoes?
3. The idea that the earth is divided into plates that have moved away from each other through time is called what?
4. The area which includes most of the earth's volcanoes and earthquakes is called what?
5. What specific major historical geologic event took place in the 600th year of Noah' life on the first day of the Genesis Flood?

Quiz #3:

1. Name the mafic volcanic minerals. Name the felsic volcanic minerals.
2. Write out the chemical symbols of the eight most abundant elements.
3. Name an example of one each of the mafic, felsic and intermediate volcanic rocks.
4. A rock consisting of bits and pieces of ash, glass and other volcanic rocks usually fused into a hard rock is called?
5. What is the word for *fire broken*?
6. A plutonic rock is referred to as what kind of rock?
7. A volcanic rock is referred to as what kind of rock?

Quiz #4:

1. The term tephra describes what kind of volcanic material?
2. What are the differences between a Hawaiian type eruption and a Plinian type eruption? Which one would be most likely to produce pyroclastic flows?
3. What two common factors influence the explosivity of an eruption?
4. Which of the following eruptions is most likely to produce rhyolite lava? Hawaiian or Pelean?
5. What is the earth's largest shield volcano?
6. What is a table mount?
7. What is a phreatic eruption?
8. Vulcanian eruptions are most likely to produce what kind of lava?

Quiz #5:

1. What types of volcanic material have been produced by the historic eruptive events mentioned in this lesson?
2. What factors, brought about by the Genesis Flood, would have produced the massive eruptive events mentioned in this lesson?
3. Why should we question the uniformitarian (secular) explanation of these historical eruptive events?

Quiz #6:

1. What are the three main parts of the atom?
2. Give an example of one of the differences between a stable and an unstable atom.
3. Why can't we depend on radiometric dating to give accurate and dependable dates for the age of rocks?
4. Cite one example of where radiometric dating has been demonstrated to be unreliable.

Quiz #7:

1. What are some of the poisonous gases that are emitted during a volcanic eruption?
2. Besides poisonous gases, name two other hazards of volcanic eruptions.

Quiz #8:

1. Name the four types of thermal features and briefly describe each one.
2. What are black and white smokers? How are they similar to geysers?
3. What are thermophiles?
4. What is travertine?
5. What is geyserite?
6. What is the main difference between a caldera and a stratovolcano?
7. What is the main type of lava produced by the Yellowstone Caldera?

Quiz #9:

1. Define the word seismic.
2. What is the difference between a tsunami and a tidal wave?
3. What is an escarpment?
4. What are the four types of faults?
5. What is a rift? How are faults and rifts related to volcanism?
6. What are seismic waves?

Quiz #10:

1. Name four geologic evidences for an Ice Age.
2. What affect would an increase in volcanism have on an Ice Age?
3. Why would it be necessary to have both warmer oceans and a cooler atmosphere in an Ice Age?
4. If the Northern Hemisphere and higher latitudes had more snow and ice during the Ice Age, what would the rest of the world have experienced? Why?

Quiz #11:

1. Name two of the lessons from Mt. St. Helens that teach us about a Biblical understanding of the Genesis Flood.
2. How did geologists arrive at an age of 40,000 years for Mt. St. Helens?
3. Where in the Biblical chronology would you put the formation of Mt. St. Helens?

Quizzes Answer Key

Quiz #1

1. James Hutton
2. "The present is the key to the past." Present geological processes are sufficient to explain the geologic history of the earth
3. God probably created the universe, but is no longer involved in its processes or maintenance
4. The amount of quartz
5. Fire broken; refers to the broken bits and pieces of volcanic rock, hot steam and gasses, and glass which hug the ground in fast-moving clouds
6. It is an index which uses past historical eruptions to measure the power of present volcanic eruptions.
7. A fissure
8. Magma
9. Secular view looks at the present geology and interprets the past without reference to the Scriptures. The Biblical view starts with the Scriptures as God's revelation and proceeds to interpret the present rock layers and landforms.

Quiz #2

1. Stratovolcano (composite volcano), cinder cone, caldera, shield, fissure, dome (may include seamounts also)
2. Lava of lava flows has low viscosity and is generally basaltic; stratovolcanoes produce andesitic lava eruptions and the lava is more viscous.
3. Plate Tectonic Theory
4. The Ring of Fire
5. The fountains of the great deep burst open.

Quiz #3

1. Amphibole, pyroxene, biotite, calcium feldspar, olivine and iron; quartz. potassium feldspar, sodium feldspar, muscovite mica
2. O, Si, Al, K, Na, Fe, Ca, Mg
3. Mafic – basalt; felsic – rhyolite; intermediate – andesite
4. Welded tuff
5. Pyroclastic
6. Coarse-grained
7. Fine-grained

Quiz #4

1. Tephra = any ejected material like ash, lapilli, tuff, pumice, bombs and blocks.
2. Hawaiian type = fluid, basaltic, calm; Plinian = explosive, massive eruptive column of ash high into the atmosphere. The Plinian eruption would be most likely to produce pyroclastic flows.
3. The magma mix and contact of hot magma with water
4. The Pelean eruption
5. Mauna Loa
6. Underwater volcano
7. Hot magma coming into contact with water causes violent steam explosions
8. Andesitic to dacitic

Quiz #5

1. Pyroclastic, rhyolite and basalt lava flows, bombs, ash and tuff
2. Hot magma and lots of water from the breaking up of the fountains of the great deep; cracks and fissures
3. It has never been nor could it ever be scientifically documented. It is built on naturalistic assumptions about the past, not scientific observation.

Quiz #6

1. Nucleus consisting of protons and neutrons and the outer shells consisting of electrons
2. Carbon-12 is a stable element. Carbon-14 is unstable; it has too many neutrons
3. It is based on improvable assumptions
4. Mt. St. Helens new dacite dome was dated thousands of years older than its 10 year age.

Quiz #7

1. Carbon dioxide, sulfur dioxide, hydrogen sulfide, arsenic
2. Lahars and pyroclastic flows

Quiz #8

1. Geyser – water spouter; hot spring – hot water, but not eruptive; mud pot – hot spring with sulfur dioxide; thermophiles convert it to useful energy and the by product is sulfuric acid which dissolves the geyserite, turning it to mud; fumarole – hot steam only.
2. A black smoker is located on the ocean bottom near fissures in the oceanic crust. They release hot gasses and metallic sulfide minerals which make them black; white smokers do the same except they release the lighter hued minerals.
3. A thermophile is a heat-loving bacterium in the hot thermal waters of volcanically active areas.
4. Travertine is the deposit produced from the dissolution of limestone rich spring water.
5. Geyserite is the deposit produced from the dissolution of rhyolite rich spring water; it is also known as sinter.
6. A caldera is a huge opening in the earth subsequent to a violent volcanic eruption; it is really a collapse of an exceptionally big volcano. Its lava is primarily rhyolitic. A stratovolcano is a high cone built on the successive eruptions and lava flows with andesitic lava.
7. Rhyolite

Quiz #9

1. To shake from an earthquake
2. A tsunami is a seismic wave generated by underwater earthquakes; a tidal wave is generated by the tides.
3. An escarpment is a break between two moving pieces of Earth.
4. Normal sip and slip fault; Reverse upward fault; lateral or strike slip fault; oblique either up or down along with lateral movement
5. A rift is a huge crack in the earth usually extending over a large land area.
6. Shock waves generated by earthquakes

Quiz #10

1. Striations; erratic boulders; cirques; glacial troughs; glacial loess; horns
2. To cool the atmosphere by an effusion of volcanic ash
3. It is one thing to cool the atmosphere, but you must also have an abundance of evaporation from warmer oceans to generate the massive amount of snowfall needed.
4. Wetter weather

Quiz #11

1. Rapidly deposited layers of sediment; rapidly cut canyons; rapidly deposited stripped trees, with some in apparent growth position – all pointing to the catastrophic, rapidly eroded Earth during the Genesis Flood
2. Through radiometric dating and comparison to the Geologic Time Table
3. It would have to be placed after the Flood sometime. Otherwise it would have been washed away.

Comprehensive Exam

Multiple Choice – choose the best possible answer

1. Andesite is a type of
- *a)* Rhyolitic lava
- *b)* Intermediate lava
- *c)* Basaltic lava
- *d)* Dacitic lava

2. The kind of lava that characterizes the Hawaiian volcanoes is
- *a)* Basaltic lava
- *b)* Pyroclastic lava
- *c)* Welded tuff
- *d)* Andesitic lava

3. Mauna Loa is what type of volcano
- *a)* Stratovolcano
- *b)* Caldera
- *c)* Cinder Cone
- *d)* Shield

4. The father of modern geology was
- *a)* Charles Darwin
- *b)* Charles Lyell
- *c)* James Hutton
- *d)* William Smith

5. Hot molten rock that extrudes onto the surface of the earth is called
- *a)* Tuff
- *b)* Lava
- *c)* Magma
- *d)* Ash

6. The idea that present processes can explain Earth history is called
- *a)* Evolution
- *b)* Big Bang
- *c)* Gradualism
- *d)* Uniformitarianism

7. The type of volcano that collapse in on itself after an initial violent explosion is called
- *a)* Caldera
- *b)* Composite Volcano
- *c)* Shield
- *d)* Fissure Volcano

8. Things like volcanic bombs and lapilli are referred to as
- *a)* Ash
- *b)* Tephra
- *c)* Tuff
- *d)* Lava flows

9. The area of the earth that includes most of the world's volcanoes is called
- *a)* Tectonic Plate
- *b)* Earthquake Zone
- *c)* Tsunami Zone
- *d)* The Ring of Fire

10. The major ingredient that determines viscosity in lava is
- *a)* Potassium Feldspar
- *b)* Darker Minerals
- *c)* Lighter Minerals
- *d)* Quartz

11. A volcano that erupts just once in its lifetime is a
- *a)* Resurgent Dome
- *b)* Cinder Cone
- *c)* Composite Volcano
- *d)* Shield Volcano

12. The major lava that erupts from Hawaiian volcanoes is
- *a)* Rhyolite lava
- *b)* Dacite lava
- *c)* Andesite lava
- *d)* Basalt lava

13. A table mount is a type of
- *a)* Mountain
- *b)* Black smoker
- *c)* Volcano
- *d)* Fissure

14. The Columbia Plateau is representative of
- *a)* Shield Volcanoes
- *b)* Seamounts
- *c)* Rhyolite Flow
- *d)* Basalt Flow

15. The type of eruption that involves hot magma and water in a violent steam explosion is a
- *a)* Strombolian eruption
- *b)* Phreatic eruption
- *c)* Vesuvian eruption
- *d)* Shield eruption

16. The type of lava most erupted from Yellowstone Caldera was
   a) Basalt
   b) Rhyolite
   c) Dacite
   d) Andesite

17. A plutonic rock is also called a
   a) Coarse grain rock
   b) Fine grain rock
   c) Pegmatite
   d) Volcanic rock

18. The word mafic stands for
   a) Iron and quartz
   b) Magnesium and olivine
   c) Quartz and magnesium
   d) Magnesium and iron

19. The word felsic stands for
   a) Quartz and sodium feldspar
   b) Feldspar and olivine
   c) Quartz and feldspar
   d) Feldspar and calcium

20. An example of an unstable element is one that has
   a) Just the right number of protons
   b) More protons than electrons
   c) More neutrons than electrons
   d) More neutrons than protons

21. A poisonous gas released during a volcanic eruption is
   a) H2O
   b) N14
   c) O2
   d) CO2

22. A rock consisting of bits and pieces of fused rock is called
   a) Welded tuff
   b) Pyroclastic ash
   c) Loess
   d) Porphyritic clasts

23. The most abundant element in the earth's crust is
   a) Oxygen
   b) Silicon
   c) Potassium
   d) Aluminum

24. An example of a mafic rock is

a) Rhyolite
b) Basalt
c) Andesite
d) Dacite

25. The type of thermal feature that erupts hot water and steam is called a
a) Hot spring
b) Thermal hot spring
c) Mud Pot
d) Geyser

26. The rock formed from the precipitation of limestone out of a hot spring is called
a) Granite
b) Stalactite
c) Geyserite
d) Travertine

27. The rock formed from the precipitation of quartz out of a thermal feature is called
a) Geyserite
b) Travertine
c) Lime deposit
d) Layered granite

28. The remnants of a glacially carved mountain are called
a) Striations
b) Erratics
c) Loess
d) Horns

29. The Geology Community ultimately relies on what tool to dictate how old the earth is
a) Radiometric Dating
b) The Geologic Timetable
c) The Evolutionary Succession of the Fossils
d) Whether lava has oxidized

30. Bacteria that can live at extreme temperatures are called
a) Extraterrestrials
b) Algae
c) Stromatolites
d) Thermophiles

Comprehensive Exam Answer Key

1. b
2. a
3. d
4. c
5. b
6. d
7. a
8. b
9. d
10.d
11.b
12.d
13.c
14.d
15.b
16.b
17.a
18.d
19.c
20.d
21.d
22.a
23.a
24.b
25.d
26.c
27.a
28.d
29.b
30.d

Credits

Lesson One:
Mount Fuji - Public Domain, https://commons.wikimedia.org/w/index.php?curid=1395736, p. 9;
James Hutton - "Sir Henry Raeburn - James Hutton, 1726 - 1797. Geologist - Google Art Project" by Henry Raeburn - uwGnf-SEw_8MjQ at Google Cultural Institute, zoom level maximum. Licensed under Public Domain via Wikimedia Commons – http://commons.wikimedia.org/wiki/File:Sir_Henry_Raeburn_-_James_Hutton,_1726_-_1797._Geologist_-_Google_Art_Project.jpg#/media/File:Sir_Henry_Raeburn_-_James_Hutton,_1726_-_1797._Geologist_-_Google_Art_Project.jpg., p. 10; Charles Lyell: "Lyell 1840". Licensed under Public Domain via Wikipedia - http://en.wikipedia.org/wiki/File:Lyell_1840.jpg#/media/File:Lyell_1840.jpg., p. 11; Rock Cycle - "Rockcycle" by Woudloper/Woodwalker ; Licensed under Public Domain via Wikimedia Commons – http://commons.wikimedia.org/wiki/File:Rockcycle.jpg#/media/File:Rockcycle.jpg., p. 12;

Lesson Two
Volcano chart: Courtesy USGS, p. 14; Volcanic fissure and lava channel: "Volcano q". Licensed under Public Domain via Commons - https://commons.wikimedia.org/wiki/File:Volcano_q.jpg#/media/File:Volcano_q.jpg, p. 14; Shield volcano: Photo by Vicki S. Nurre, p. 15; Shield Volcano vs. Composite Volcano courtesy USGS, p. 15; Shield volcano vs. Composite Volcano: Public Domain, USGS, p. 15; Craters of the moon: Photo by Patrick Nurre, p. 15; SP Crater: "SP Crater' by USGS - http://wrgis.wr.usgs.gov/fact-sheet/fs017-01/Transferred from en.wikipedia to Commons by Hike395 using CommonsHelper.. Licensed under Public Domain via Commons - https://commons.wikimedia.org/wiki/File:SP_Crater.jpg#/media/File:SP_Crater.jpg, p. 15; Cinder Cone on Mauna Kea: Patrick Nurre, p.16; Mt. Fuji: "Mount Fuji from meadow" by skyseeker - https://www.flickr.com/photos/skyseeker/11078938913. Licensed under CC BY 2.0 via Commons - https://commons.wikimedia.org/wiki/File:Mount_Fuji_from_meadow.jpg#/media/File:Mount_Fuji_from_meadow.jpg, p. 16; Mt. Hood: "Mount Hood 3233" by Walter Siegmund (talk) - Own work. Licensed under CC BY-SA 3.0 via Commons - https://commons.wikimedia.org/wiki/File:Mount_Hood_3233.JPG#/media/File:Mount_Hood_3233.JPG, p. 16; Dome volcano: "Volcán Chaitén-Sam Beebe-Ecotrust" by Sam Beebe - Volcán Chaitén. Licensed under CC BY-SA 2.0 via Commons - https://commons.wikimedia.org/wiki/File:Volc%C3%A1n_Chait%C3%A9n-Sam_Beebe-Ecotrust.jpg#/media/File:Volc%C3%A1n_Chait%C3%A9n-Sam_Beebe-Ecotrust.jpg, p. 16; Mt. St. Helens: "MSH82 st helens plume from harrys ridge 05-19-82" by Lyn Topinka - CVO Photo Archive Mount St. Helens, Washington Before, During, and After 18 May 1980, Licensed under Public Domain via Commons - https://commons.wikimedia.org/wiki/File:MSH82_st_helens_plume_from_harrys_ridge_05-19-82.jpg#/media/File:MSH82_st_helens_plume_from_harrys_ridge_05-19-82.jpg, p. 17; Composite drawing of Mt. St. Helens Dome: By Surachit - self-made SVG, based on public domain USGS image [1], CC BY-SA 3.0, https://commons.wikimedia.org/w/index.php?curid=2569731, p. 17;
360 view of Mt. St. Helens: "Mount St Helens Summit Pano II" by Farwestern Photo by Gregg M. Erickson - Own work. Licensed under CC BY 3.0 via Commons - https://commons.wikimedia.org/wiki/File:Mount_St_Helens_Summit_Pano_II.jpg#/media/File:Mount_St_Helens_Summit_Pano_II.jpg, p. 17; Yellowstone Caldera: Map courtesy of United States National Park Service, nps.gov, public domain, p. 18; Diagram of laccolith: by en:User:Erimus, User:Stannered - en:Image:Laccolith.JPG, Public Domain, p. 19;
Laccolith: Found at https://en.wikipedia.org/wiki/Laccolith#/media/File:Limestone_Butte_Montana_Laccolith.jpg, public domain, p. 19; Round Mountain: Found at https://www.flickr.com/photos/39422575@N02/2557011961/, public domain, p. 19; Lone Mountain: Photo by Vicki Nurre, p. 19;
Square Butte: By Mike Cline - Own work, CC BY-SA 4.0, https://commons.wikimedia.org/w/index.php?curid=40798388, p. 19; Devil's Tower 1900: By Nathaniel H. Darton - U.S. Geological Survey Photographic Library, Public Domain, https://commons.wikimedia.org/w/index.php?curid=11648305, p. 21;
Devil's Tower: NPS, public domain, p. 21; La Sal Mountains: CC BY-SA 3.0, https://commons.wikimedia.org/w/index.php?curid=375034, p. 21; Ring of fire map: Courtesy of USGS, p. 22; Mid-Atlantic Ridge: By originally uploaded on the English wikipedia: 14:51, 21 October 2003 . . JamesDay (Talk / contribs) 200×415 (21,177 bytes) (mid-Atlantic ridge map) - [1], Public Domain, https://commons.wikimedia.org/w/index.php?curid=1623126, p. 23; Mid-Atlantic rift diagram: "Iceland Mid-Atlantic Ridge Fig16" by USGS - pubs.usgs.gov. Licensed under Public Domain via Commons - https://commons.wikimedia.org/wiki/File:Iceland_Mid-Atlantic_Ridge_Fig16.gif#/media/File:Iceland_Mid-Atlantic_Ridge_Fig16.gif, p. 23; Rift in Iceland: (photo) By Diego Delso, CC BY-SA 4.0, https://commons.wikimedia.org/w/index.php?curid=37290290, p. 23; Parts of a volcano: "Volcano scheme" by MesserWoland - own work created in Inkscape. Licensed under CC BY-SA 3.0 via Commons - https://commons.wikimedia.org/wiki/File:Volcano_scheme.svg#/media/File:Volcano_scheme.svg, p. 24;

Lesson Three
Abundant minerals in the earth: Vicki S. Nurre, p. 26; Mafic Minerals: Patrick Nurre, p. 27; Felsic Minerals: Patrick Nurre, p. 27; Dark-colored Lavas: Photos by Vicki S. Nurre, p. 28-30; Light-colored volcanic rocks: Photos by Vicki Nurre, p. 31-33;

Intermediate volcanic rocks: p. 33-34; Porphyritic Volcanic Rocks: Photos by Vicki Nurre, p. 34; Pyroclastic Volcanic Rocks: Photos by Vicki Nurre, p. 35; Pele's tears, Pele's hair: Photos by Vicki Nurre, p. 35; Basalt Bombs: Photos by Vicki Nurre, p. 36; Ash: Photo by Vicki Nurre, p. 36; Pumice: Photos by Vicki Nurre, p. 36; Lava Chart: Vicki Nurre, p. 37;

Lesson Four
Hawaiian eruption diagram: "Hawaiian Eruption-numbers" by © Sémhur / Wikimedia Commons. Licensed under FAL via Commons - https://commons.wikimedia.org/wiki/File:Hawaiian_Eruption-numbers.svg#/media/File:Hawaiian_Eruption-numbers.svg, p. 40; Pahoehoe lava: Photo by Vicki Nurre Lava: Photo by Vicki Nurre, p. 40; Strombolian eruption diagram: "Strombolian Eruption-numbers" by © Sémhur / Wikimedia Commons. Licensed under FAL via Commons - https://commons.wikimedia.org/wiki/File:Strombolian_Eruption-numbers.svg#/media/File:Strombolian_Eruption-numbers.svg, p. 40; Photo of Strombolian eruption: "Stromboli Eruption" by Wolfgangbeyer at the German language Wikipedia. Licensed under CC BY-SA 3.0 via Commons - https://commons.wikimedia.org/wiki/File:Stromboli_Eruption.jpg#/media/File:Stromboli_Eruption.jpg, p. 40; Volcanian eruption diagram: "Vulcanian Eruption-numbers" by Sémhur (talk) - Own work, from the document about volcanism on the Portail sur la prévention des risques majeurs (web portal about the prevention of the major risks) of the Ministère français de l'Ecologie, du Développement et de l'Aménagement durables (french Minister of the Ecology, Environment and Sustainable Development).. Licensed under FAL via Commons - https://commons.wikimedia.org/wiki/File:Vulcanian_Eruption-numbers.svg#/media/File:Vulcanian_Eruption-numbers.svg, p. 41; Photo of Volcanian eruption: "Tavurvur volcano 2" by Taro Taylor - originally posted to Flickr as Eruption. Licensed under CC BY 2.0 via Commons - https://commons.wikimedia.org/wiki/File:Tavurvur_volcano_2.jpg#/media/File:Tavurvur_volcano_2.jpg, p. 41; Pelean eruption diagram: "Pelean Eruption-numbers" by © Sémhur / Wikimedia Commons. Licensed under FAL via Commons - https://commons.wikimedia.org/wiki/File:Pelean_Eruption-numbers.svg#/media/File:Pelean_Eruption-numbers.svg, p. 41; Mayon volcano: "Pyroclastic flows at Mayon Volcano" by C.G. Newhall - http://volcanoes.usgs.gov/Imgs/Jpg/Mayon/32923351-020_caption.html. Licensed under Public Domain via Commons - https://commons.wikimedia.org/wiki/File:Pyroclastic_flows_at_Mayon_Volcano.jpg#/media/File:Pyroclastic_flows_at_Mayon_Volcano.jpg, p. 41; Plinian eruption diagram: "Plinian Eruption-numbers" by © Sémhur / Wikimedia Commons. Licensed under FAL via Commons - https://commons.wikimedia.org/wiki/File:Plinian_Eruption-numbers.svg#/media/File:Plinian_Eruption-numbers.svg, p. 42; Redoubt volcano: "MtRedoubtedit1". Licensed under Public Domain via Commons - https://commons.wikimedia.org/wiki/File:MtRedoubtedit1.jpg#/media/File:MtRedoubtedit1.jpg, p. 42; Surtseyan eruption diagram: "Surtseyan Eruption-numbers" by © Sémhur / Wikimedia Commons. Licensed under FAL via Commons - https://commons.wikimedia.org/wiki/File:Surtseyan_Eruption-numbers.svg#/media/File:Surtseyan_Eruption-numbers.svg, p. 43; Photo Island of Surtsey: "Surtsey eruption 1963". Licensed under Public Domain via Commons - https://commons.wikimedia.org/wiki/File:Surtsey_eruption_1963.jpg#/media/File:Surtsey_eruption_1963.jpg, p. 43; Submarine eruption diagram: "Submarine Eruption-numbers" by © Sémhur / Wikimedia Commons. Licensed under FAL via Commons - https://commons.wikimedia.org/wiki/File:Submarine_Eruption-numbers.svg#/media/File:Submarine_Eruption-numbers.svg, p. 43; Loi'hi: "Loihi 3d" by John Smith and Brooks Bays - (Original text: "National Science Foundation / National Oceanic and Atmospheric Administration / The Hawaii Undersea Research Laboratory (HURL) http://www.soest.hawaii.edu/HURL/images/loihi_3d.gif http://www.soest.hawaii.edu/HURL/hurl_loihi.html). Licensed under Public Domain via Commons - https://commons.wikimedia.org/wiki/File:Loihi_3d.gif#/media/File:Loihi_3d.gif, p. 43; Phreatic eruption diagram: "Phreatic Eruption-numbers" by Sémhur · ✉ · ✍ · - own work / travail personnel.Inspired by the document about volcanism from the Portail sur la prévention des risques majeurs (web portal about the prevention of the major risks) of the Ministère français de l'Ecologie, du Développement et de l'Aménagement durables (french Minister of the Ecology, Environment and Sustainable Development).. Licensed under CC BY-SA 3.0 via Commons - https://commons.wikimedia.org/wiki/File:Phreatic_Eruption-numbers.svg#/media/File:Phreatic_Eruption-numbers.svg, p. 44; Photo Mt. St. Helens: "MSH80 eruption mount st helens 05-18-80" by Austin Post, USGS - Huge tif converted to jpeg and caption from USGS Mount St. Helens, Washington May 18, 1980 Eruption Images. Licensed under Public Domain via Commons - https://commons.wikimedia.org/wiki/File:MSH80_eruption_mount_st_helens_05-18-80.jpg#/media/File:MSH80_eruption_mount_st_helens_05-18-80.jpg
Photo Mt. St. Helens: "Phreatic St Helens" by USGS - United States Geological Survey - http://volcanoes.usgs.gov/Products/Pglossary/HydroVolcEruption.html. Licensed under Public Domain via Commons - .https://commons.wikimedia.org/wiki/File:Phreatic_St._Helens.jpg#/media/File:Phreatic_St._Helens.jpg, p. 44;
Relative Explosiveness: Image by Vicki Nurre, information found at https://intlpollution.commons.gc.cuny.edu/volcanic-pollution/, p. 45; Volcanic Explosivity Index: By chris 論 - Originally from http://volcanoes.usgs.gov/Products/Pglossary/vei.htmlAccessed September 29, 2005, Public Domain, https://commons.wikimedia.org/w/index.php?curid=3935253, p. 45; Honey: public domain, p. 46;
Oil: By SIGAUS - www.sigaus.es, CC BY-SA 4.0, https://commons.wikimedia.org/w/index.php?curid=54361166, p. 46;
Lava: Public domain, By Tari Noelani Mattox,[1] USGS geologist[2][3] - https://web.archive.org/web/20070102035046/http://volcanoes.usgs.gov/Products/Pglossary/pahoehoe_ropy.html, Public Domain, https://commons.wikimedia.org/w/index.php?curid=700082, p. 46; Map of lavas on Big Island: Public domain, USGS, p. 46; Kalapana buried by lava flow: By Rolfcosar - Own work, CC BY-SA 4.0, https://commons.wikimedia.org/w/index.php?curid=68925724, p. 46; Volcano type as function of gas content: Vicki Nurre, p. 47; Classification and Flow Characteristics of Volcanic Rocks: Based on information found at USGS.gov., by Vicki Nurre, p. 48; Viscosity example: Photo by Vicki Nurre, p. 49;

Lesson Five

Volcanic ejecta comparison: Based on information found at https://geology.com/stories/13/volcanic-explosivity-index/, by Vicki Nurre, p. 50; Columbia Flood Basalts Map: By Victor Camp and Martin Ross - Obtained from author., CC BY-SA 3.0, https://commons.wikimedia.org/w/index.php?curid=18060697, p. 51; Columbia Flood Basalts: Photo by Vicki Nurre, p. 51; Map, Deccan Traps: Vicki Nurre, p. 51; Deccan Traps: Photo By Nicholas (Nichalp) - Own work, touched up in Photoshop, CC BY-SA 2.5, https://commons.wikimedia.org/w/index.php?curid=1806702, p. 51; Basalt flows map: Vicki Nurre, based on information found at Based on info found at https://blogs.scientificamerican.com/history-of-geology/large-igneous-provinces-and-mass-extinctions/, p. 51; Yellowstone supervolcano map: Public domain, NPS, p. 52; Pilot Peak and Index Peak: Photo by Vicki Nurre, p. 52; Ash beds: By en:User:G. Thomas - EN:WP http://en.wikipedia.org/wiki/Image:WheelerGACO.jpg, Public Domain, https://commons.wikimedia.org/w/index.php?curid=2442811, p. 53; Mt. Mazama: By Semionk at the English language Wikipedia, CC BY-SA 3.0, https://commons.wikimedia.org/w/index.php?curid=2745808, p. 53; Mt. Tambora: - Image and and English description: Mount Tambora Volcano, Sumbawa Island, Indonesia, NASA Earth Observatory. 2nd version: Transferred from de.wikipedia to Commons.; originally from https://eol.jsc.nasa.gov/SearchPhotos/photo.pl?mission=ISS020&roll=E&frame=06563, Public Domain, https://commons.wikimedia.org/w/index.php?curid=7336599, p. 54; Giant's Causeway: Public domain, found at https://www.flickr.com/photos/133997242@N03/26041067755/, p. 54; Tamu Massif: Public Domain By IODP - http://www.sci-news.com/othersciences/geophysics/science-tamu-massif-largest-volcano-pacific-01366.html, Public Domain, https://commons.wikimedia.org/w/index.php?curid=33053895, p. 55;

Lesson Six

Periodic Table of the Elements: Free to use without attribution, p. 57; Radioactivity: Image by Vicki Nurre, p. 58; Carbon 12 and Carbon 14 representation: Vicki Nurre, p. 58; Geologic Time Scale (2): Vicki Nurre, p. 60; Unconformity: Photo by Vicki Nurre, p. 61; Periodic Table of the Elements: Free to use without attribution, p. 62;

Lesson Seven

Mt. St. Helens: Public domain, USGS, p. 63; Mt. St. Helens diagram: https://volcanoes.usgs.gov/volcanoes/st_helens/st_helens_gallery_27.html, USGS, p. 64; pH scale: Vicki Nurre, p.65; Statue: Public domain By Slick - Own work, CC0, https://commons.wikimedia.org/w/index.php?curid=22257663, p. 65; Aftermath of Lahar: By Robin Holcomb, U.S. Geological Survey - Archived source link, Public Domain, https://commons.wikimedia.org/w/index.php?curid=5883394, p. 65; Ashfall: CC BY-SA 2.0, https://commons.wikimedia.org/w/index.php?curid=126389, p. 66; Mayon Volcano: Mayon Volcano: Public Domain By C.G. Newhall - http://volcanoes.usgs.gov/Imgs/Jpg/Mayon/32923351-020_caption.html, Public Domain, https://commons.wikimedia.org/w/index.php?curid=246427, p. 66; Mt. St. Helens trees: https://volcanoes.usgs.gov/volcanoes/st_helens/st_helens_gallery_23.html, p. 67; Lava flow: Lava flow during rift eruption, Krafla, Iceland Public domain By Michael Ryan, U.S. Geological Survey. - earthquakes krafla, Public Domain, https://commons.wikimedia.org/w/index.php?curid=2575167, p. 67; Volcanic hazards: Public Domain, By U.S. Geological Survey - http://volcanoes.usgs.gov/hazards/, Public Domain, https://commons.wikimedia.org/w/index.php?curid=9771298, p. 68; Depiction of acid rain: USGS, public domain, p. 68; Volcanic features: Public domain, contains only material that came from the USGS, p. 68; Mt. St. Helens, after eruption: Public domain, US Forest Service 2007, p. 69; Mt. St. Helens recovery: Free at pixabay no license required, no attribution required, p. 69;

Lesson Eight

Bison: Photo by Karen, Dennis, used by permission, p. 71; Lone Star Geyser: Photo by Vicki Nurre, p. 71; Lower Falls: Photo by Vicki Nurre, p. 71; Mt. Mazama caldera collapse: By U.S. GEOLOGICAL SURVEY and the NATIONAL PARK SERVICE - http://pubs.usgs.gov/fs/2002/fs092-02/ (saved as PNG), Public Domain, https://commons.wikimedia.org/w/index.php?curid=667211, p. 71; Basalt flow: Photo by Vicki Nurre, p. 72; Rhyolite flow: Photo by Patrick Nurre, p. 72; Rhyolite obsidian flow: Photo by Vicki Nurre, p. 72; Volcanic tuff: Photos by Vicki Nurre, p. 72; Volcanic ash: https://volcanoes.usgs.gov/volcanoes/st_helens/st_helens_gallery_27.html, p. 73; Fountain Geyser: Public domain, National Park Service, p. 73; Beehive Geyser: Photo by Patrick Nurre, p. 73; How a geyser works: Public domain, National Park Service, p. 74; Sinter/geyserite (3): Photos by Vicki Nurre, p. 74; Mammoth Terraces: Photo by Karen Dennis, p. 75; Travertine: Photo by Vicki Nurre, p. 75; Fumarole: Photos by Vicki Nurre, p. 75; Fumarole: Photo by Vicki Nurre p. 75; Fumarole: Photo by Karen Dennis, used by permission, p. 75; Hot spring (2): Photo by Vicki Nurre, p. 75; Mud pot: Photo by Vicki Nurre, p. 76; Mud pot: By David Monniaux - Own work, CC BY-SA 3.0, https://commons.wikimedia.org/w/index.php?curid=1327117, p. 76; Thermophiles (4): photos by Vicki Nurre, p. 76; Travertine feature: Photo by Karen Dennis, used by permission, p. 76; Black smoker: Found at https://oceanservice.noaa.gov/facts/vents.html Public Domain, NOAA, p. 77; White smoker: Public Domain, NOAA By NOAA - http://oceanexplorer.noaa.gov/explorations/04fire/logs/hirez/champagne_vent_hirez.jpg, Public Domain, https://commons.wikimedia.org/w/index.php?curid=3597762, p. 77;

Lesson Nine

Earthquake pictures (in order): Public domain By Guasinay50 - Own work, Public Domain, https://commons.wikimedia.org/w/index.php?curid=5182696, p. 79; 2nd picture public domain, p. 79; 3rd pic public domain By Roger Bilham - https://www.ngdc.noaa.gov/hazardimages/event/show/24, Public Domain, https://commons.wikimedia.org/w/index.php?curid=66305961, p. 79; 4th picture public domain By H.G. Wilshire, U.S.

Geological Survey - Photo by H. G. Wilshire for U.S. Geological Survey.High res version on line at : [1]Lower resolution version: Original image here: http://pubs.usgs.gov/dds/dds-29/web_pages/oakland.html, Public Domain, https://commons.wikimedia.org/w/index.php?curid=1417555, p. 79; Earthquake epicenters: By NASA, DTAM project team - http://denali.gsfc.nasa.gov/dtam/seismic/, Public Domain, https://commons.wikimedia.org/w/index.php?curid=35429, p. 80; Seismograph: By Yamaguchi 先生, CC BY-SA 3.0, https://commons.wikimedia.org/w/index.php?curid=1089235, p. 76; Types of earthquakes: Three basic fault types https://soundwaves.usgs.gov/2009/11/fieldwork2.html, p. 81; Hebgen Earthquake (2): By R.B. Colton - USGS Photographic Library - ID. Colton, R.B. 658ct (Direct link), Public Domain, https://commons.wikimedia.org/w/index.php?curid=13361155; USGS public domain; public domain By J.R. Stacy - USGS, Public Domain, https://commons.wikimedia.org/w/index.php?curid=47678408, p. 81; Jordan Rift Valley (2): https://creativecommons.org/licenses/by-sa/4.0/deed.en By Zairon, p. 82; Interior of the earth: Image by Vicki Nurre, p. 82; Epicenter diagram: By AnsateSam Hocevar (original author; this is a derivative work)User:TFerenczy create SVG version; cs translationUser:NikNaks es translationUser:Lies Van Rompaey nl translationUser:Rostik2520004 ru translationUser:Ата uk translation - Epicenter.png:, CC BY-SA 1.0, https://commons.wikimedia.org/w/index.php?curid=32716741, p. 83; Tsunami damage (2): Public domain, p. 84;

Lesson Ten

Ice Sheet: USGS found at https://pubs.usgs.gov/gip/continents/, p. 87; Erratics: Photo by Vicki Nurre, p. 88; Glacial striations (2): Photos by Vicki Nurre, p. 88; Glacial horns and cirques (2): Photos by Vicki Nurre, p. 88; Glacial trough: Photo by Vicki Nurre, p. 89; Atmospheric result of volcanoes: Public domain https://www.nasa.gov/vision/earth/lookingatearth/volcano_nile.html, p. 89;

Lesson Eleven:

Map of Mt. St. Helens: Public Domain, USGS, p. 92; Mt. St. Helens before and after: Public domain, USGS, p. 93; Canyon: Public domain, USGS, p. 94; Mt. St. Helens aftermath: Public domain, USGS, p. 94; South Dakota Badlands: Public domain, USGS, p. 94; Ash strata: Public domain, USGS, p. 95; Debris avalanche deposit (2): Public domain, USGS, p. 95; Fossil trees diagram, Yellowstone: Public domain, NPS, found at https://www.nps.gov/parkhistory/online_books/yell/knowlton/sec1.htm, p. 96; Specimen Ridge, Yellowstone National Park: Photo by John Meyer, used with permission, p. 96; Log mat: Public domain, USGS, p. 96; Mt St. Helens today: Public domain, USGS, p. 96;

Appendices

Plate tectonics: By Map:USGSDescription:Scott Nash - File:Tectonic plates.png, Public Domain, https://commons.wikimedia.org/w/index.php?curid=535201, p. 98; Sea floor spreading: By NASA/GSFC/Robert Simmon - http://earthobservatory.nasa.gov/Study/Tectonics/tectonics_3.html, Public Domain, https://commons.wikimedia.org/w/index.php?curid=358818, p. 99; Iceland rift: By Pmarshal - Own work, Public Domain, https://commons.wikimedia.org/w/index.php?curid=4430821, p. 99; Map aerial view: By NASA/MODIS - Jacques Descloitres, MODIS Land Rapid Response Team - http://visibleearth.nasa.gov/view_rec.php?id=1939, Public Domain, https://commons.wikimedia.org/w/index.php?curid=659187, p. 99; Map: By Bourrichon - fr:Bourrichon with English translations, additions, and corrections by Ketone16 (partly following Yuri Koryakov) - Own work ;Topographic data from the NASA Shuttle Radar Topography Mission (SRTM3 v.2) (public domain) edited with 3DEM, reprojected in UTM with GDAL (GDAL), and vectorized with Inkscape ; UTM projection ; WGS84 datum ; shaded relief (composite image of N-W, W and N lightning positions) ;Reference used for the additional data :* Rivers, cities, railroads, routes, bathymetry : Demis add-on for World Wind (see the approval e-mail and the Demis forum) ;* Borders and coast : World data bank II ;Approximate scale of topographic data : 1:1,463,000 ;Note : The shaded relief is a raster image embedded in the SVG file., CC BY-SA 4.0, https://commons.wikimedia.org/w/index.php?curid=50309413, p. 99;

As a youngster, Patrick Nurre spent many Saturdays rockhounding near the Big Horn River in Montana, leading to a lifelong study of geology. Today, Patrick is a popular speaker on geology and worldview, and teaches classes and leads field trips, including his Yellowstone Creation Adventure. In 2005, he started Northwest Treasures, which is devoted to designing geology kits for schools. He conducts numerous field trips each year in Washington State to such places as the Olympic Peninsula, Mt. Rainier, Mount St. Helens, the Channeled Scablands, Mt. Baker and Whidbey Island. In addition, he also gives field trips to the volcano loop of Oregon and California, Mt. Hood volcanic area (Oregon), the eastern badlands of Montana and Yellowstone National Park. He is the author of seventeen geology textbooks and is a Certified Biblical Creation Ministry Professional. He and his wife, Vicki, live in the Seattle, Washington area, and have three wonderful children and two grandchildren.

If you would like to contact Patrick about speaking or field trips: northwestexpedition@msn.com

For a list of speaking topics: NorthwestRockAndFossil.com

Made in United States
North Haven, CT
25 September 2026

19145838R00C69